TOO DUMB TO METER

Follies, Fiascoes, Dead Ends and Duds on the U.S. Road to Atomic Energy

KENNEDY MAIZE

ISBN: 1466420529
ISBN-13: 9781466420526

Library of Congress Control Number: 2011960973
CreateSpace, North Charleston, South Carolina

TABLE OF CONTENTS

Masters of the Universe

On September 16, 1954, a dapper and double-breasted U.S. Atomic Energy Commission Chairman, Lewis Strauss, stepped to the microphone at the Waldorf Astoria hotel in New York City. He was addressing a dinner meeting of the National Association of Science Writers. An ebullient Strauss bragged,

> Transmutation of the elements—unlimited power, ability to investigate the working of living cells by tracer atoms, the secret of photosynthesis about to be uncovered—these and a host of other results all in fifteen short years. It is not too much to expect that our children will enjoy in their homes electrical energy too cheap to meter—will know of great periodic regional famines in the world only as matters of history—will travel effortlessly over the seas and under them and through the air with a minimum of danger and at great speeds—and will experience a lifespan far longer than ours, as disease yields and man comes to understand what causes him to age. This is the forecast for an age of peace.

It didn't quite work out that way. Much of the story of the atom focuses on the well-known course of development of newer, bigger, stronger bombs, and of the birthing problems and maturation of civilian nuclear power plants.

Whereas most histories of technology catalog success, this book details failure: explosive, expensive, repeated failure. This is a less well known story, but often more interesting and more amusing. It also serves as a cautionary account of the perils of government hubris, public hysteria, and centralized planning gone wild: misguided policy, misunderstood history, misapplied engineering, and mistaken economics. This book brings to light some of the things that went wrong—often terribly wrong—from conception through failed implementation. It is a tale of the stubborn and mistaken belief in the ability of big science, big engineering, and big government money to solve any technical problem.

This story begins not with the well-known history of the Manhattan Project and its intrepid bomb builders, but with what came next—immediately after August 1945.

1. The Madness of Nukes

Most Americans reacted joyously to the initial news of the atomic bombs falling on Japan in the summer of 1945, as the vast destruction spread by the atom spelled the end of the Empire of the Rising Sun. Whatever these mysterious bombs were, they did the job—and most were glad. Sen. Brien McMahon, a young Connecticut Democrat who would seal his brief place in history by becoming a chief architect of the postwar Atomic Energy Commission, was fond of saying that the bombing of Hiroshima was the greatest event in world history since the birth of Jesus Christ.

Yet, there was also a twinge of guilt in the public sentiment after the first atom bombs fell, particularly as the scope of the devastation in the two Japanese cities became known. Writer John Hersey captured the ambiguity beautifully in an article titled "Hiroshima," published in the *New Yorker* in August 1946, a year after the U.S. bomb destroyed that metropolis.

Hersey's 36,000-word article, which occupied the entire edition of the magazine and was immediately published as a book, personalized the effects of the atomic bomb in spare, calm language that made the horror of nuclear war accessible to any reader. It had a profound impact on the way many people viewed atomic energy for decades to come.

Hersey—who had won a Pulitzer Prize for fiction the year before with *A Bell for Adano*, a tale of the U.S. military occupation of a town in Italy—described six survivors of the nuclear inferno in Japan and how it changed their lives. When the article appeared, the magazine's editors began the edition with this introduction:

> TO OUR READERS the *New Yorker* this week devotes its entire editorial space to an article on the almost complete obliteration of a city by one atomic bomb, and what happened to the people of that city. It does so in the conviction that few of us have yet comprehended the all but incredible destructive power of this weapon, and that everyone might well take time to consider the terrible implications of its use. - The Editors."

"Hiroshima" was a publishing sensation. The magazine quickly sold out on newsstands (at fifteen cents an issue), and copies were soon being scalped to collectors for ten dollars and more. Reprint requests poured in to the *New Yorker*, and Knopf produced a book that hit the stores in October. The Book of the Month Club distributed copies to its members for free. It became one of the most influential books of the last half of the twentieth century.[1]

Soon after, many Americans began to wonder incredulously at those two terrible pieces of blast and fire that fell out of the sky. Just what were they? Clearly, a new force had been unleashed, one that even most well-educated Americans didn't understand and couldn't quite comprehend.

Time magazine, in its July 1, 1946, cover story on Albert Einstein and the bomb (the third time Einstein had appeared on the magazine's cover[2]), captured that feeling of combined awe and befuddlement, writing in its signature style:

> It is typical of the dilemma of this civilization that masses of men humbly accept the fact of Einstein's genius, but only a handful understand in what it consists. They have heard that, in his Special and his

---

1 The book was still in print fifty years after the original article appeared in the New Yorker.

2 Images of Einstein and his family appeared on the cover of *Time* eight times between 1929 and 2007. He was Time's "Man of the Century" in the final 1999 issue.

General Theories of Relativity, Einstein finally explained the form and the nature of the physical universe and the laws governing it. They cannot understand his explanation. To a small elite of mathematicians and physicists, the score of equations in which Einstein embodied his picture of the universe and its functioning are as concrete as a kitchen table. To the layman they are as staggering as to be told, when he is straining to make out the smudge which is all he can see of the great cluster in the constellation Hercules, that the faint light that strikes his eye left its source 34,000 years ago.

Time concluded that most people would never understand much about Einstein's theories—the fundamental ideas behind nuclear energy—beyond this limerick:

There was a young lady called Bright,
Who could travel much faster than light;
She went out one day,
In a relative way,
And came back the previous night.

Much of this wonder and incredulity grew out of the secrecy of the atomic adventure. As it emerged after the war, the story of the Manhattan Project was a revelation. The largest engineering and military production effort in history had occurred in the United States over a period of nearly five years, completely under the unknowing noses of the American public. It had all been hidden in veils of secrecy, and now the story was beginning to unfold.

Just how secret? The vice president of the United States, Harry S. Truman, didn't know about the atomic bomb until after his boss, President Roosevelt, died in April 1945. Four months before the two bombs fell on Hiroshima and Nagasaki and only three months before the successful Trinity test at Alamogordo in the New Mexico wilderness, Truman listened in astonishment when he got his first briefing. With only sketchy understanding of what had been going on, he soon had to make the fateful decision to let the atomic demon loose on the world.

3

Gen. Leslie Groves, the career military man who ran the project, describes in his 1962 memoir, *Now It Can Be Told*, how the military bamboozled Congress on the program through 1943. The enormous atomic bomb project was buried in a series of War Department sub-accounts within the impenetrable military budget. Even the War Department bureaucrats responsible for allocating the money were largely in the dark. Groves talks about a "bad moment in late 1943," when Rep. Albert J. Engel (R-MI) got wind of a major construction project in the Tennessee woods at a place called, Oak Ridge. Engel wanted to make a trip to Oak Ridge to see what was going on. "In reply," Grove writes, "he was told that this work was highly secret, and that the information he wanted could not be given to him; eventually, he was persuaded to forget his contemplated visit."

Ironically, while Groves and his atomic bomb babysitters were able to keep most of Congress, the vice president, most of the war and foreign policy bureaucracy, and all of the American people in the dark, that didn't work with our sometimes ally and long-term adversary throughout the twentieth century—the former Soviet Union. Stalin and his spymasters knew a great deal about the secret endeavor and were quickly able to demonstrate their own explosive prowess with atomic science.

The Manhattan Engineering District—the cover name for the atomic bomb program—created what Groves called the "country's greatest single scientific success." It also created a couple of enduring myths. First, the bomb builders' success fed the notion that large, government-directed and -funded scientific and engineering programs can overcome almost any technical, political, or social obstacles. The later success of the Apollo moon project reaffirmed that belief. The residue of that notion of government-driven science can be seen in the subsequent history of public policy in the twentieth century, and today. The Nixon administration's hopelessly hubristic War on Cancer in the early 1970s exemplified the lingering paradigm of the Manhattan Project, as did the Carter administration's support for creating a giant synthetic fuels industry in the 1980s,[3] which turned into a colossal flop. Today, in the bowels of the Department of Energy, the

---

3 Carter, a Naval Academy graduate and a nuclear engineer, was part of Adm. Hyman Rickover's elite Navy nuclear program until he had to leave the service early to take over his family's peanut business.

Manhattan mentality remains, fueling research and development in such areas as: how to capture and stuff into the ground carbon dioxide from coal-fired power plants; how to economically turn sunlight directly into electricity; and how to midwife a new generation of nuclear power technologies.

Almost immediately after the atomic bombs fell on Hiroshima and Nagasaki, the United States went nuts over nukes. The shadowy world of the atom, rumored in technical journals and occasional "gee-whiz" newspaper or magazine article in the pre-war press, burst onto the scene in 1945. The result was a tidal wave of enthusiasm for anything and everything atomic. Anyone associated with the atom was a rock star[4]; the atom was the future of the universe; and the United States seemed the master of that universe. The nation was enthralled by hyper-optimistic notions about what the atom could do, beyond blowing up enemy cities and spreading radioactive fallout around the globe. Trains, boats, and planes would be atom-powered. Tiny atomic reactors would sit in our basements and heat our houses. Government would beat into peaceful plowshares the most terrible sword humankind had ever developed.

Writer Daniel Ford, who covered nuclear energy for the *New Yorker* twenty-five years after Hersey, described in his book, *Cult of the Atom*, a "general euphoria" about atomic energy. Ford linked that feeling to the undercurrent of guilt left from the bombings. "Instead of reflecting on the horrors visited upon Hiroshima and Nagasaki or on whether the bombs should have been used in the first place," Ford wrote, "news reports helped to alleviate the nation's feelings of repulsion and guilt by focusing public attention on the more congenial aspects of 'the new force.'"

---

4 Between Einstein's appearance on July 1, 1946, cover and 1961, men (entirely men) associated with atomic power appeared repeatedly on *Time* magazine's cover. The list included: Harvard scientist and presidential science advisor James B. Conant (twice); first Atomic Energy Commission chairman David Lilienthal; seminal atomic scientist and father of the fission bomb J. Robert Oppenheimer (twice); AEC chairman Gordon Dean; AEC chairman and failed nominee as Eisenhower's Commerce Secretary Lewis Strauss (twice); Navy nuclear chief Adm. Hyman Rickover; AEC chairman and Nobel laureate Willard Libby; scientist and political advisor Lee DuBridge; seminal nuclear scientist and father of the fusion bomb Edward Teller (twice); and Nobel laureate and AEC chairman Glenn Seaborg.

A mere two weeks after bombs fell on Japan, *Newsweek* gushed that "even the most conservative scientists and industrialists were willing to outline a civilization which would make the comic-strip prophecies of Buck Rogers look obsolete." In December of 1945, just four months after the attack on Japan, *Popular Science* magazine proclaimed in a cover story headline "We CAN Harvest the Atom." The article went on to say, "You will soon see mobile engines running on U235, and cities heated by steam from stationary graphite piles." A 1953 *Look* magazine article by Gordon Dean, one of the original members of the postwar Atomic Energy Commission, was titled "Atomic Miracles We Will See."

Over the years, the hyperbole rolled on.

The military, the civilian government, and the popular press[5] touted nuclear power as a panacea to many of the military and domestic problems that faced the nation. Bizarre notions of the prospects of nuclear energy for enriching civilian life took hold in these influential circles. Take the family sedan. Ford Motor Co. in 1958 created a concept car, called the "Nucleon," designed to be powered by a tiny nuclear reactor. It existed, or course, only on drawing paper and a 3/8-scale clay mockup. But the Ford Nucleon is evidence[6] of how the atom was the dream of the age in the 1950s and 1960s.

Even comic strip characters were enlisted in the army of atomic acolytes. One was Dagwood Bumstead, the harried and harassed, suburban, sandwich-loving salary-man who was the ever flappable hero of the *Dagwood and Blondie* strip. The strip has been a fixture on newspaper comic pages for more than 70 years and was the prototype for generations of television sitcoms from the Honeymooners to Ozzie and Harriet and Ricky and Lucy to Mad Men.

In September 1948, *Popular Science* magazine carried an episode titled "Learn How Dagwood Splits the Atom," a piece of pure propaganda, but with considerable educational content. The following year, King Features,

---

5 Shamefully, the press at the time complied and cooperated with the government's overt manipulation. William Laurence, the Pulitzer Prize-winning science reporter for the *New York Times*, was on the payroll of the Manhattan Project, at the same time he was being paid by the newspaper, with the knowledge and agreement of the editors. He wrote the Army press release announcing the successful Trinity test in New Mexico.

6 It is on display today at the Ford museum in Dearborn, Michigan.

the syndicate that distributed the strip to newspapers around the country, published the Dagwood atomic energy strip as a free-standing comic book. The Dagwood comic book featured a foreword by well-known journalist Bob Considine and a formal endorsement from Gen. Leslie Groves. Although it is not clearly stated in the document nor is there any evidence to support the conclusion, it is hard to believe that the book did not have Atomic Energy Commission funding.

In 1951, the A.C. Gilbert Company of Fair Haven, Connecticut, maker of toys for budding scientists and engineers[7] came out with the "U-238 Atomic Energy Lab," a briefcase-sized case full of radiation goodies for inquisitive kids. The fifty-dollar kit (very expensive for the day) included four different types of uranium ore, a Geiger counter for measuring radiation, a spinthariscope for seeing atoms split naturally, and a miniature cloud chamber for tracking different sub-atomic particles. The lucky child also received a government-issued pamphlet titled "Prospecting for Uranium" aimed at aiding would-be prospectors (with the possibility of a ten thousand dollar reward from the government for a good discovery of uranium ore), and a copy of the Dagwood comic book.

While the bombs were bad, the atom was good. That was the message the government was pitching in the aftermath of the war. The popular president, Dwight Eisenhower, touted what he dubbed Atoms for Peace in 1954 (partly to overcome widespread feelings that the atomic scientists and bureaucrats were not delivering on their hyperbolic claims), and the Post Office issued a three-cent Atoms for Peace first-class stamp in 1955. Some 133 million stamps came off government printing presses.

Even Disney, the juggernaut of popular culture, got into the act of promoting the beneficial atom. Working with publisher Simon and Schuster in 1956, Disney produced the large-format book *Our Friend the Atom*, written by expatriate German physicist Heinz Haber.[8] Disney artists illus-

---

7 Among Gilbert's toys were American Flyer trains (better than Lionel, but less popular), Erector sets, chemistry sets that generations of (mostly) boys used to make stink bombs, and the atomic energy kit. Gilbert's scientific toys and the company's tale are on display at the Eli Whitney Museum and Workshop in New Haven, Connecticut.

8 American science benefited in many ways from the brutal rise of Nazism in Germany and its spread across Europe in the 1930s and after the war. While there were plenty

7

tiated the work. In the foreword, Walt Disney himself (or a ghostwriter) wrote, "Atomic science began as a positive, creating thought. It has created modern science with its many benefits for mankind. In this sense our book tries to make it clear to you that we can indeed look upon the atom as our friend."

In 1954, New York publishing house Grosset & Dunlap relaunched a series of books aimed at ten- to fourteen-year-old boys intrigued with technology. The books were the second generation of Tom Swift kids' science novels, named the Tom Swift Jr. line. Both the original Tom Swift books, which began in 1910 and saw distribution until 1941, and the post-war iteration of the 1950s through 1971, were aimed at similar generations of young readers, primarily boys, hooked on technology.

The putative author of the second run of books was Victor Appleton II: a concocted moniker for a group of writers working on a rigid formula that carried the series through a dozen books. Their inspiration was the phenomenal advancement of nuclear and military science that characterized the end of the war, as the public became drunk with the prospects of science and technology in the aftermath of the Manhattan Project.

When Tom Swift Jr. stepped onto the fictional stage, everything seemed possible.

The original Tom Swift series—also written, under the Victor Appleton pseudonym, by a collection of authors writing to formula—articulated a similar reverence for scientific advancement and its purported solutions. Tom Sr. invented the picture telephone, vertical takeoff aircraft, and a giant military tank—all prescient, though not all of his inventions eventually saw the light of day.

Tom Sr. also gave us the delightful Tom Swifties puns, which remain a parlor game among some aging pop literature raconteurs. In the game,

---

of homegrown scientific giants in the United States, such as E.O. Lawrence, J. Robert Oppenheimer, and Ted Taylor, the flow of scientific talent from Europe was astonishing: Einstein, Teller, Ulam, Bethe, Fermi, Szilard, Gamow, von Braun, Ley, and, Haber. Born in 1913, Haber earned a doctorate in physics in Berlin and became a Luftwaffe aviator until 1942. He emigrated to the United States in 1946, where he worked on space medicine projects. While teaching physics at UCLA, he became the leading scientific consultant to Walt Disney productions.

one is asked to come up with adverbial, adjectival, or other puns with Tom quotes, mimicking the original Tom. For example:

- "Who would want to steal modern art?" asked Tom abstractedly.
- "Fire!" yelled Tom alarmingly.
- "It's a unit of electric current," said Tom amply.
- "Why invade Iraq?" Tom said ironically.
- "Another batch of shells for me!" Tom clamored.
- "George W. Bush?" asked a dumbfounded Tom.

Tom Swift Jr.'s escapades continued the tradition and exemplified the technological optimism of the nuclear world after the end of World War II. Tom was the son of the original, who by that time had made a fortune from his inventions.

An ebullient eighteen-year-old, Tom Jr. and his friends, relying on their own inventiveness, his father's advice, and the money from his father's engineering enterprises, were able to conceive and develop a series of new technologies, without the use of government funds and in astonishingly short time.

These inventions inevitably saved the nation from the nefarious plots of foreign governments. Our adversaries in the Swift books invariably were bogeymen from Eastern Europe or South America. They were dark-skinned, secretive, and motivated by hatred of the United States and a desire to supplant American power with their own.

All this played into the fears of the day. In the wake of the war, Soviet power advanced to conquer central and eastern Europe. Communism captured China. "Who lost China," was the refrain of right-wing Congressional Republicans, as if Harry Truman and the Democrats—not U.S. support for the corrupt government of Chiang Kai-shek—which led to Mao Tse-tung and his agrarian Communists were responsible.

But while the alleged traitors in our government, proclaimed by Republican Sens. Joseph McCarthy of Wisconsin and John Bricker of Ohio and others in both parties, were said to be selling the nation down the drain, technology would rescue us. No one was as good as the United States at turning basic science into useful weapons, goods, and services. That the godless Commies had managed to develop their own nuclear weapons (which they thankfully never used) was solely a result of espionage and theft. This was the gospel of the friendly atom circa 1954.

The Tom Swift books represented the technological illusions of the post-war period. Tom[9] was lanky, sporting a blond crew-cut, and almost always wearing a T-shirt with blue and white horizontal stripes, and blue jeans. True to formula, he had a heroic sidekick, Bud Barclay, who was darker, shorter, and stockier than Tom. A good athlete, Bud was not nearly as intellectually gifted as Tom (who was?). He often came to Tom's rescue when the hero was captured by the enemy. Also in sync with the formula, Tom had a comic sidekick, Charles "Chow" Winkler, a former cowboy chuck-wagon cook who had become the Emeril Legasse of Swift Enterprises. He was prone to loud clothes and bizarre outbursts such as "brand my space biscuits" that are as charming as the earlier Tom Swifties. The infectious optimism of Tom Swift and his crew carried over to government policy makers, such as Lewis Strauss (pronounced "Straws"). A former shoe sales-man, he became a wildly successful and rich investment banker. Appointed to the newly created Atomic Energy Commission by President Truman in 1946 and President Eisenhower's choice as chairman in 1954, Strauss's optimism characterized the times.

Strauss also symbolized the shift from the military to civilian control over the power of the atom in the United States. The Manhattan Engineering Division became the Atomic Energy Commission after a politically contentious battle, which in the end created a formal structure outside the military for the development of nuclear energy. The new structure, however, did little to dilute the power of the military over nuclear energy. The organizational chart changed, but the mind-sets of the masters of the atom remained militaristic.

---

9 Tom started as an eighteen-year-old and never aged a day during the seventeen-year run of the series.

2. Manhattan Transfer

Following the end of World War II, the victorious United States took as one of its first tasks asserting control over the atom: both the physical force unleashed by breaking its strong bonds; and the institutional, business, bureaucratic, and political forces unleashed by mastering that physical force. The government had to turn the military rule of the jungle, which produced the bomb, into a rule of law, which would dictate the future refinement and production of nuclear power and weaponry. It is no accident that the official Atomic Energy Commission history of this immediate postwar period, written by George T. Mazuzan and J. Samuel Walker, published in 1984, is titled "Controlling the Atom, the Beginnings of Nuclear Regulation 1946–1962." A volume covering the earlier period, written by Richard G. Hewlett and Oscar E. Anderson Jr., was aptly titled "The New World, 1939–1946."

The issue facing the United States after the war was how to put civilians in control of what had been an entirely military program, without sapping the strengths of the military success. It was a bit like the task of taming a wild horse, where the ideal is to harness the strength and spirit of the mustang into a model saddle pony. Not an easy task.

Two characteristics dominated the United States government when it came to atomic energy during and after World War II: overwhelming reliance on secrecy and overweening technological hubris. The traits were well-earned, as demonstrated by the success of the Manhattan Project. That endeavor and the resultant weapons not only ended the war against Japan but also altered the balance of power in the postwar period, and led to the Cold War nuclear equilibrium between the U.S. and the Soviet Union that lasted for more than 50 years. The Cold War competition to build more and better bombs and faster, smarter and more deadly delivery and defense systems may have effectively spent the Soviet Union into oblivion.

The United States nuclear weapons program from 1942–1946 was the most secret large-scale military project in history. Most of the details, including the most mundane, were not made known until a major push by the Clinton administration in the early 1990s declassified much of the documentation. The pursuit of invisibility dominated the nuclear endeavor from its days as an entirely military mission through its transformation into a civilian enterprise, to the point of obsession.

The postwar period was also intensely, radioactively partisan, which affected the debate over what to do about the atom. Democrats had been in control in Washington since Roosevelt's 1932 presidential election. The president had won unprecedented third (1940) and fourth (1944) terms. Vice President Truman continued the Democratic hegemony when he succeeded to the highest office in 1945. The Democratic Party controlled Congress, as well as the White House, during that thirteen-year period.

By 1946, American voters were understandably restive. The war was over, and it was time to return to a more normal existence. Truman tried, but failed, to continue wartime rationing of strategic commodities. Industries had tired of wartime production quotas and demands from Washington. Car makers wanted to make cars, not tanks; and consumers wanted to buy cars and new tires and gasoline without needing government-allocated coupons or bypassing rationing for the black market.

Republicans sensed that they had an opportunity to make great political gains as the nation came out of its war environment. One of the first points of contention between the parties—a harbinger of political fights to come—was how to organize the nation's nuclear energy research and development program. Republicans, with a tradition of fealty to military leadership and concepts and a desire to draw a favorable line in the shifting political sand, started pushing a civilian regime that retained most of the wartime military controls, including the dogma of intense inscrutability. For the most part, the military was perfectly happy with an arrangement that put civilians in charge of a weak institution that continued to respond primarily to military needs.[10]

The Republican view of the postwar atomic venture resonated with many conservative, military-oriented Democrats, splitting the majority party. The Democratic Party also had a tradition that valued government control over market forces, a holdover from the Depression- and New Deal-era approach to creating large, government-controlled institutions to solve social problems, such as the Tennessee Valley Authority.

But the majority party was riven. Arrayed against the Southern-dominated right wing of the Democrats were the internationalists, with a vision drawn in part from their fealty to the ideas of their icons, Woodrow Wilson and his failed League of Nations, and Roosevelt and his variant on the theme, the United Nations. Also dividing the Democrats were those in the Truman White House who wanted as much authority and flexibility located in the executive branch as possible and congressional Democrats, who wanted to exercise greater control over their rivals at the other end of Pennsylvania Avenue.

Complicating the politics were the Manhattan Project scientists, many of whom felt drawn back to the academic life, where they exercised control over their research agendas without any direction from politicians and bureaucrats. Many of the scientists, Oppenheimer for example, were also staunch internationalists who wanted to see atomic energy put out of the reach of nationalist aims, whether those of Washington or Moscow.

The Manhattan Project flew in the face of the ethical and epistemological traditions of science in general and physics in particular. Traditionally, science had relied on open discussion, give-and-take, argument, and sharing experimental data and research results. But that ran counter to the needs of the military to compartmentalize data, restrict access to research, and stifle scientific debate. Many feared that open science would arm our enemies, initially the Germans, but increasingly the Soviet Union.

---

10 Republicans also wielded the loyalty cudgel against Truman's tottering Democratic administration, but lost control of the wildly dysfunctional and alcoholic McCarthy, who sowed his own doom when he attacked the Republican Eisenhower administration in 1953 and 1954.

In the Manhattan Project, the Army created an open, but extremely limited, arena for physicists and other scientists to hash out their diverse views, in remote and completely controlled environments in the New Mexico wilderness, the wilds of Tennessee, the desert of eastern Washington, and elsewhere. Among the scientists who lived and worked in the endeavor, free scientific debate was the order of the day. But Often, they couldn't even discuss what they were working on with their families. The code of secrecy grated on many.

While the Manhattan Project scientists were few in number, they possessed considerable political power and were justly admired by the general public; they owned "the knowledge," the expertise on how to make the bombs. The nation could ill afford to lose the efforts of the atomic scientists—among the finest physicists, chemists, mathematicians, metallurgists and technicians ever assembled.

The result of this brew of conflicting interests, desires, and political imperatives was the Atomic Energy Act of 1946. A fundamentally compromised and flawed piece of legislation, the 1946 law created an unaccountable administrative structure that would dominate the U.S. atomic energy program for thirty years. Its legacy is felt today. Produced by a special congressional committee assembled to paper over the schisms that divided Congress and the Truman administration, the act created a carefully-balanced structure unique in the government. It possessed enormous responsibility and power with little external oversight and disregard for the Constitutional imperative of separation of powers.

For the first eight years, the AEC functioned essentially as a faux military agency. Congress mandated continued government ownership of nuclear materials, technology, and know-how. There was no civilian nuclear power industry in the United States in the first decade following the victory in World War II. Twenty years later, there were still only a handful of nuclear power plants in the United States. But billions had been spent on military and civilian schemes—many of them harebrained.

In turning the military program into nominally civilian hands, Congress created myriad institutional and bureaucratic problems. It built a structure that proved perfect for intrigue, politics, pork-barrel spending, and low-level (and sometimes higher) bureaucratic warfare among competing power centers. The Atomic Energy Commission; the White House and its Bureau of the Budget; the Pentagon; industry lobbyists; and the Joint Committee on Atomic Energy all participated in a bureaucratic cotillion of program authorizations, funding, schedules, and allocations of resources. It provided an unfortunate model for Congressional behavior continuing to the present.

While the 1946 law established the AEC to manage the weapons program, the military remained intimately involved. Military decisions drove the shape of the AEC program, in terms of the kinds of bombs, the numbers of warheads, and even major decisions on nuclear technology. Adm. Hyman Rickover, for example, the father of the nuclear submarine, kept both his military rank and a position as civilian head of an office at the AEC. The same dual military-civilian arrangement was established in the AEC–Air Force nuclear bomber program and other ventures.

A five-member commission, appointed by the president, governed the AEC. The chairman came to be considered the face of atomic energy in the United States. The AEC's general manager, typically out of the limelight, ran the day-to-day activities of the agency. Depending on the strengths and wishes of the chairman, AEC general managers sometimes were important powers unto themselves under the light-handed and often heedless direction of the chairman and commission, sometimes partners linking the AEC staff and its policy and political overseers, sometimes simply spear-carriers for the commission and Congress.

Offering independent scientific advice to the commission, a nine-member General Advisory Committee consisted of presidentially appointed civilians and important scientists charged with giving the commissioners— who often had little or no scientific background—the best technical advice available. The 1946 law gave the GAC the task to "advise the Commission on scientific and technical matters relating to materials, production, and research and development, to be composed of nine members, who shall be appointed from civilian life by the President." As a practical matter, the AEC staff typically gave the White House the names of nominees to the GAC, effectively choosing who would oversee their operations. The GAC often turned out to be a critic and sometimes a scold and goad to AEC projects, particularly when Oppenheimer was its chairman.

A Military Liaison Committee (MLC)—created by Sen. Arthur Vandenberg, the respected Michigan Republican,[11] during the congressional debate over the 1946 act—balanced the civilian views of the GAC. While dedicated to civilian control of nuclear energy, Vandenberg understood that the commission would continue to have close ties to the military, including developing, testing, and producing the nuclear weapons stockpile for the nation as well as research and development for a new generation of weapons. He believed the military, specifically the Army Chief of Staff, should continue to have input at the commission on military issues. Groves, who ran the Manhattan Project with intense personal interest and care, and who opposed turning the endeavor into a civilian satrapy, lobbied for a continued major military role.

Congress conceded that the MLC should provide military advice to the commission. The secretaries of War and the Navy, and later the Air Force, had the responsibility to name the military committee members. The first MLC was made up entirely of men chosen by Groves, positioning himself as first head of the committee. Groves never really accepted the idea that civilians should run his military offspring, which had done so well under his parentage.

Congress also kept military personnel in place in the civilian bureaucracy through the creation of an AEC staff office, the Director of Military Applications. Congress earmarked this slot for a military officer, and the office would effectively run the commission's day-to-day military activities, reporting to the commission through the general manager.

David Lilienthal proved the dominant figure of the early days of the AEC. Named, by Truman, the first chairman of the new, civilian agency, his confirmation hearing before the nine Senate members of the Joint Committee on Atomic Energy—created specifically to consider the 1946 Atomic Energy Act—turned into a classic partisan and ideological brawl that presaged some of the Red Scare years of the 1950s. It was not a good omen for the AEC's future as an independent agency.

Republicans saw Lilienthal's nomination as an opportunity to score hits on the Democrats and the Truman administration. Some of the GOP

---

11 Vandenberg became chairman of the Senate Foreign Relations Committee, serving from 1947–1948, after the Republicans won the 1946 off-year elections.

members of the Senate were carrying jurisdictional water for Groves, who remained a foe of the new AEC. Others wanted to use Lilienthal's nomination as an opportunity to blast the creation and performance of the Tennessee Valley Authority, one of the New Deal's signature accomplishments, which Lilienthal headed. Republican orthodoxy regarded TVA as at least the perfidious camel's socialist nose under the tent of private-sector electric utilities. Sen. Styles Bridges, a New Hampshire Republican, criticized Lilienthal for leading the TVA, "a social experiment, which is a wide departure from the American system of private ownership of property." Ohio Republican Bob Taft, whom the press branded "Mr. Republican," and who was a potential Republican presidential candidate in 1948, opposed Lilienthal as "temperamentally unfitted to head any important executive agency in a democratic government and too 'soft' on issues connected with communism and Soviet Russia."

The fiercest opposition to Lilienthal came from a fellow Democrat, albeit a conservative one, Sen. Kenneth McKellar of Tennessee. Having joined the senate in 1917 and serving as the senate president *pro tempore* until the GOP electoral success of 1946, McKellar loathed Lilienthal, largely because Lilienthal—as TVA board member since 1935 and chairman starting in 1941—had resisted McKellar's efforts to turn the TVA into a patronage honeypot. One history of the period describes McKellar as possessing "a mind warped by age and a smoldering hatred."[12] In Lilienthal's AEC confirmation hearings, McKellar resurrected bogus charges, rejected a decade before, that Lilienthal was a member of a TVA communist cell. McKellar also reportedly raised anti-Semitic arguments against Lilienthal behind the scenes.

The hearings so disturbed Lilienthal that, according to historians Richard Hewlett and Francis Duncan, he seriously considered withdrawing his name from the nomination. But Clark Clifford in the White House prevailed on Lilienthal, arguing that most of what had been occurring was just for show. Neither Taft nor McKellar, no matter how vitriolic their rhetoric, were members of the joint committee (they appeared at the hearings as a

---

12 Lilienthal once told a reporter that his favorite actor was the late comic W.C. Fields, "and not because he looks like Kenneth McKellar."

17

matter of Senate courtesy and custom), and would have to save their fire for a floor vote, a more difficult proposition.

Lilienthal and Clifford assiduously worked the committee and firmed up the votes. Neither McKellar nor Taft seemed to have much influence. Republican Bourke Hickenlooper of Iowa, the chairman, was a solid vote for Lilienthal, as was the revered Vandenberg. On March 10, 1947, the Senate component of the joint committee voted 8–1[13] to approve the nomination. The confirmation hearings had begun on January 27.

While Lilienthal's opponents, led by Taft, tried to mount a Senate floor fight against the AEC nomination, that rear-guard action collapsed after Vandenberg, who succeeded McKellar as president *pro tempore* in the Eightieth Congress, delivered a strong rebuttal to his fellow Republicans on Lilienthal's fitness to lead the AEC. The vote on Lilienthal was 52–38 in favor.

What Lilienthal inherited from a reluctant Leslie Groves represented an unprecedented administrative challenge. By the end of the war, the Manhattan Project was a vast enterprise, run from a remote site near Oak Ridge, Tennessee, which also housed a giant factory for separating uranium and drew large amounts of electricity from the government's Tennessee Valley Authority. The nuclear weapons empire included New Mexico's Los Alamos; a huge weapons factory in the desert on the Columbia River near Hanford, Washington; an assembly plant near Denver, Colorado; a plutonium fabrication factory in Texas; and a reactor testing station in the eastern Idaho desert.

Time magazine in 1947 described the scope of the endeavor that the AEC took over from the Manhattan Engineering District: "It was a maze of contracts, documents, libraries, factories, laboratories, whole towns. It was shrouded and obscured in the bright fog of military security. It was jealously guarded by the Army's Major General Leslie Groves, then the District's

---

13 The "no" vote came from Republican Sen. John Bricker of Ohio. Voting for the Lilienthal nomination were Republicans Hickenlooper, Vandenberg, Eugene Millikan of Colorado, and William Knowland of California. All four Democrats—Brien McMahon of Connecticut, Richard Russell of Georgia, Edwin Johnson of Colorado and Tom Connally of Texas—voted for Lilienthal.

chief, now a disgruntled member of the Military Liaison Committee, embittered by the lack of kudos for his wartime stewardship."

On the lists of land, men, and equipment the Army transferred to the AEC in 1947, were thirty-seven installations in nineteen states and Canada. The manpower included 254 military officers, 1,688 enlisted men, 3,950 government civil service workers, and some 38,000 contractor employees. At least, that's what the Army listed on the material it provided to the AEC. *Time* noted, "Groves feared that civilian control would mean dissolving the security screen. AEC asked for a complete inventory; the Army refused it."

In any case, it was an enormous undertaking. This mind-boggling venture represented an investment of over $2.2 billion, and the agency's first-year budget was $300 million—a vast sum by 1946 standards.

Lilienthal was well-suited to get the new agency moving under civilian control. He had shown in taking control of the newly-created TVA that he could build a massive federal agency from scratch. Of course, the new problem he faced was not starting from nothing, but taking an enormous existing structure and getting it headed in a different direction. Lilienthal relished the task. He described himself as a "craftsman in public affairs."

Time commented on Lilienthal: "To men like Bob Taft, he is the symbol of the New Deal, of Big Government, of hostility to business. To his friends, he is a public servant of the highest order. His ability is cited by his friends as an argument in his favor, by his enemies as a proof of his danger. On one point everyone is agreed: Lilienthal, who loves horses, is a hard rider of men and ideas."

Born in a small Illinois town in 1899 to Jewish immigrants from the Austro-Hungarian empire, Lilienthal was largely raised in Valparaiso, Indiana, where his father was a small-town, marginally successful merchant. Lilienthal graduated from De Pauw University (a school associated with the Methodist Church) in Greencastle, Indiana, where he was school newspaper editor and twice elected student body president. He then attended Harvard Law, where he came under the tutelage of legendary lawyer Felix Frankfurter, who helped his career along the way.

Lilienthal settled into a career in law in Chicago, where he built a reputation as a fine lawyer and established a successful practice, which *Time* said was bringing him twenty thousand dollars a year[14]. In 1931, Wisconsin Republican Gov. Phil La Follette named him to the five thousand dollar–per year job on the state public service commission, which regulated utilities. In 1933, Roosevelt named him to the three-member TVA board. He became chairman in 1941, where he was serving when Truman picked him for the AEC chairmanship (a job paying $15,000 annually).

Despite the partisan attack against Lilienthal and his success with two Democratic presidents, he was not a traditional Democratic operative. He began his political life as a Progressive Republican, when he received the appointment from La Follett. Lilienthal described himself as an "independent" who loathed conventional politicians. As he gained experience regulating private businesses and running a larger government enterprise, Lilienthal's view evolved such that his biographer refers to him as a "statist" but not as a partisan.

Truman, whose partisan juices often ran blood red and true blue, did not view the new atomic agency as a potential pot of patronage. The president paid no attention to party background when he picked the first commissioners for the new agency. Lewis Strauss was a well-known Republican who had been Herbert Hoover's personal secretary after World War I (and had a distinguished Navy career in World War II). William Waymack, editor of the Des Moines *Register and Tribune* newspaper, was a Republican and deputy chairman of the Federal Reserve of Chicago when Truman named him to the AEC. Sumner Pike of Maine, another Wall Street success story, had been a Republican member of the Securities and Exchange Commission, appointed by Roosevelt. Only nuclear physicist Robert Bacher, the youngest member of the commission at forty-one years old, identified himself as a Democrat. But Bacher had no political background to speak of, although he had advised Bernard Baruch on technical matters at the United Nations Atomic Energy Commission while serving on the Cornell faculty.

Of the five initial members of the AEC, only two, Lilienthal and Strauss, would significantly affect the future course of the agency. Lilienthal, who served until February 15, 1950, viewed himself primarily as a public

---

14 Lilienthal probably overstated that amount considerably, as his biographer puts the figure at about $13,000 per annum, still a substantial income for the second year of the Great Depression.

administrator, a bureaucratic fixer who valued executive power and administrative efficiency. The assessment by biographer Steven M. Neuse of Lilienthal's approach to management and administration at the TVA fit as well for his time at the AEC: "In spite of *thinking* administratively, Lilienthal had little inclination for, or interest in, traditional management. In 1937 he admitted being bored with detail and frustrated by slowness, both traits antithetical to day-to-day managerial practice."

Lilienthal's lasting legacy on the AEC was to get the organization upright, on its feet, and marching forward. He delegated the day-to-day operations of the atomic bureaucracy to an able hands-on administrator, Carroll Wilson, who served as general manager from 1946 to mid-1950, coinciding with Lilienthal's tenure. Lilienthal was what one of his TVA subordinates aptly described as "a rhetorical administrative leader." He and Wilson got the commission stumbling forward; where it went and what happened when it got there was often out of Lilienthal's control—and that of the Truman administration—in part because of an additional governing structure Congress created in 1946.

3. Micro-mismanagement by Committee

During the frenzy to manage atomic power after the war, Congress created an executive branch agency that threatened to be too independent, too powerful, and too isolated from the rest of government. Compounding their errors, perhaps in recognition of what they had created, the solons also built an edifice to insert their own power into the action on a daily basis. This proved to be a major mistake—blurring the lines between executive and legislative authority—causing no end of problems for the nation's nascent atomic energy venture.

To protect its interests in the postwar jockeying over atomic energy, Congress in the 1946 law established the Joint Committee on Atomic Energy (JCAE)[15], made up of eighteen members, nine from the Senate and nine from the House. Its mission, as stated in the law, was to "make continuing studies of the activities of the Atomic Energy Commission (AEC) and of problems relating to the development, use, and control of atomic energy." The move to erect a special committee to keep Congress involved in atomic energy reflected, in part, ambiguity about the powers that had been given to the AEC. During the debate over the legislation, Rep. Clare Booth Luce, a Connecticut Republican[16], commented that the representatives were "torn between a distaste for the vast dictatorial domestic powers

---

15 The idea of the joint committee grew out of the fight that Vandenberg won in Congress over the Military Liaison Committee in considering the 1946 law. Vandenberg wanted to clip the military's political wings and persuaded Congress to create a special, House–Senate committee to consider the atomic energy legislation, which morphed into the JCAE in the final bill.

16 She was married to publisher Henry Luce, founder of *Time*, *Fortune*, *Life* and *Sports Illustrated* magazines.

[the legislation] confers on the five-man commission, and our fears that without it we shall endanger our national security in a troubled world."

The legislation gave the joint committee—the only congressional committee established by law, not by the internal rules of the House and Senate—"exclusive jurisdiction" over atomic power. One analysis concluded, "While the JCAE was certainly a committee of Congress, in many ways it did not resemble a congressional committee, or at least few other congressional committees before or since, serving as almost a unicameral legislature within the bicameral Congress."

In the normal order in Congress, various committees generally compete with each other, bringing greater debate and diversity of views and interests to consideration of legislative issues. In today's House, for example, while the Ways and Means Committee has jurisdiction over taxes, other committees also have concurrent authority over aspects of tax policy, tempering the power of the tax panel. It's messy, but it usually works. Congress, with the Atomic Energy Act, made an end-run around traditional structures and power thereby empowering an atomic autocracy.

The 1946 law also set up a political fight between the conventional appropriations committees, which had evolved into major power centers in Congress, and the new group on the block, the JCAE. In 1951, the Senate essentially ceded the power of the purse to the joint committee, providing that three members of the joint committee would be *ex officio* member of the appropriations committee for all matters atomic. That didn't happen in the House, where a stormy relationship between the nuclear power barons and the legacy powers of the appropriators prevailed.

Because Congress—which controlled the federal government's purse—gave the joint committee extraordinary authority over the AEC, the committee quickly became the political sun around which the other atomic institutions revolved. The JCAE was, in its time, the most powerful committee in Congress—arguably the most powerful in congressional history. It was also one of the most powerful institutions of government in Washington, often eclipsing the executive branch. These (mostly) guys[17]

---

17 Throughout its history from 1946 through 1976, only one woman, Rep. Luce, served on the JCAE. She served in 1946, after the creation of the committee, and did not seek reelection that year. She was first elected to the House of Representatives in 1942

had radioactive muscles, and they flexed them often—and often on behalf of half-baked, irrational nuclear schemes.

Examining the unique powers of the committee, one of the histories of the AEC notes, "Moreover, it was the only joint committee of Congress authorized to receive proposed legislation and recommend it to the Congress." In the wacky world of Washington, that's a powerful proposition: a joint committee can write bills that largely trump the work of the other substantive committees in the House and Senate.

What's more, the JCAE developed such expertise, and based its decisions on secret testimony, which it could reveal and conceal at will, that it was able to dictate national policy to Congress and the executive branch. Also, when it was adding to the powers of the committee in 1951, Congress gave it the authority to wield a legislative veto. The committee could review proposed actions of the AEC, the White House, or other executive branch agencies, including the military, and nullify them[18]. This set up monumental battles between both ends of Pennsylvania Avenue, most often won by the joint committee.

The AEC quickly realized where the real power on atomic policy in the United States could be found, and used the joint committee to protect the agency against moves by the White House to implement administration policy that the AEC found threatening, particularly on budget matters. A symbiotic and sycophantic relationship soon developed between the AEC and the joint committee.

The intimate relationship between the congressional overseers and the agency it oversaw caused some heartburn, even on the committee. California Republican Craig Hosmer, who served on the joint committee, and who

---

and reelected in 1944. Similarly, only two women ever served on the Atomic Energy Commission. The first was Mary I. Bunting, a microbiologist and president of Radcliffe College. She took a leave of absence for one year to serve as Johnson Administration AEC commissioner from mid-1964 to mid-1965. The second woman on the AEC was Dixie Lee Ray, a marine biologist, appointed in 1972 by President Nixon and named committee chair in 1973, succeeding James R. Schlesinger. She was the final head of the AEC, serving until 1975. She was elected Democratic governor of Washington in 1976.

18 In 1983, the U.S. Supreme Court found the legislative veto unconstitutional, in the case of *Immigration and Naturalization Service v. Chadra.*

had been a young AEC lawyer earlier in his professional life, was troubled by the close relationship. In a 1960 speech on the House floor, he said: "I sometimes feel that as between members of the Joint Committee and members of the Atomic Energy Commission, there exists only a shadowy and blurred understanding of which policy matters are to be decided by the committee and which by the commission, even sometimes as to what are matters of policy and what are matters of administration. This needs clarification by force of law rather than force of personalities and customs."

Also, committee staff—who wielded extraordinary power—often moved over to the AEC, where they again exercised exceptional power due to their back channel relations with the JCAE. There was similar movement in the other direction, from the commission to the committee. James Ramey's career exemplified this revolving atomic door. In 1946, Lilienthal brought Ramey—at the time, a young lawyer on the staff of the Tennessee Valley Authority—with him to Washington as an assistant general counsel. Ramey then transferred to the AEC's Chicago office, where he worked with Adm. Rickover on the contract with Westinghouse Electric Corp. for the first nuclear submarine, the Nautilus.

In 1956, Ramey moved to the joint committee, where he was executive director, the most important staff job on the committee. It was a plum job in many ways, including pay. The joint committee was not bound by staff pay directives for other congressional committees, or by the special pay schedule that applied to the AEC (which was already higher than that of the rest of the civil service). Ramey became a powerful and influential presence, particularly on matters related to the AEC's military programs and its dealings with the Pentagon. He helped develop the information on the presence of Soviet missiles in Cuba that led to the 1962 Cuban Missile Crisis.

In 1962, President Kennedy nominated Ramey as an AEC commissioner, where he served from September 1963 to July 1973. During that period, most of it served under AEC chairman Glenn Seaborg, Ramey emerged as a major power at the AEC. Seaborg, a Nobel Prize–winning nuclear chemist, was largely a hands-off administrator during his ten years as chairman, preferring to make speeches and presentations on the wonders of the atom. Ramey stepped into that vacuum. His 2010 obituary in the *Washington Post* said, "He was described in a 1974 *New York Times* article as the 'single most influential member of the commission in the past decade,' who for many years was the 'power behind the throne' of the AEC's chairmen."

On the JCAE, no more than five of the nine members from each house could be from the same party; the leadership rotated by chamber and party control of Congress. Members of the committee tended to have safe electoral seats, so they spent years and decades on the committee. The membership of the JCAE was remarkably stable. Over the years, the same names dominated the committee. From the Senate: Democrats Richard Russell of Georgia; Clinton Anderson of New Mexico; John Pastore of Rhode Island; and Republican Bourke Hickenlooper of Iowa regularly served on the committee. During the first eight years of the committee's life, control of Congress shifted back-and-forth between the two parties, and Hickenlooper moved repeatedly from ranking minority to lead Republican. He served as chairman of the committee in the Eightieth Congress (1947–1948) and the Eighty-third Congress (1953–1954). Pastore was a three-time chairman (1962–1964, 1967–1968, and 1970–1972).

On the House side, two men dominated the joint committee: Democrats Chet Holifield of California and Mel Price of Illinois. Holifield served in the House from 1943 through 1974, becoming a nuclear lobbyist for General Electric after his congressional service. Under an informal agreement that alternated the chairmanship of the joint committee between the House and Senate, Holifield chaired the committee three times (1960–1961, 1965–1977, 1969–1970)[19]. A friendly and garrulous cigar-chomping former sports writer from East St. Louis, Price chaired the joint committee only once, serving from 1973 through 1974. First elected to Congress in 1944 after Army service, Price was also a long-serving member of the House Armed Services Committee and combined his interest in atomic energy with his position on military affairs. Price became armed forces chairman in 1975, following his service on the joint committee. House Democrats purged him from the armed services committee in 1985, believing his age

---

19 In a somewhat bizarre circumstance, Holifield, while JCAE chairman in 1970, also served that year as chairman of the AEC's Military Liaison Committee. His experience before entering Congress was making and selling men's clothing. Too young for World War I service and too old for World War II, he had no military experience.

and his unwavering support for all things military rendered him no longer
fit to lead the committee.

While often critical of the AEC and its programs, all of the members
of the JCAE were enthusiastic about the big prospects for nuclear energy.
A Congressional Research Service analysis of the joint committee written
in 2004 described the committee's "unswerving dedication to the development of nuclear power." The piece goes on to note: "The members of the
JCAE had a pro-nuclear power philosophy; definite objectives which they
believed should be achieved; and specific policies which they worked to
have implemented." When the members of the joint committee were angry
with the AEC, it generally meant they didn't believe the agency was moving fast enough. Partisanship was not a major factor in the JCAE's legislative life—pork was. The committee took great care to assure that spending
on AEC programs put money into their states and congressional districts.

Industry lobbied the joint committee when it thought the AEC wasn't
serving their needs or when it wanted to mold AEC policies. If an administration (of either party) wanted to redirect or shut down a failing AEC
program, the commission invariably enlisted the JCAE on its side, along
with industry. The joint committee represented the strongest arm of an
iron triangle of interests that included the AEC, industry, and Congress.
The administration, decidedly not part of that arrangement, was often its
enemy and occasionally its foil.

The concept of the joint committee and its unprecedented powers
has prompted some admiration over the years and recommendations that
it be duplicated for other matters of national priority. Most recently, the
National Commission on Terrorist Attacks upon the United States (the
so-called 9/11 Commission) looked fondly on the JCAE and its approach
to governance. The 9/11 Commission's July 2004 final report called for
"centralizing and strengthening congressional oversight of intelligence and
homeland security issues," with the joint committee as a model. The report
said, "Either Congress should create a joint committee for intelligence,
using the Joint Atomic Energy Committee as its model, or it should create House and Senate committees with combined authorizing and appropriations powers." Fortunately for the cause of limited government and
diversity of policy views in the process of running a government, Congress

ignored the 9/11 committee on this recommendation (and most others, as well).

As the nation's views of nuclear power soured in the mid-1970s, mostly as a result of under-delivery on the promises made in the preceding decades, Congress completely restructured the government's atomic enterprise. In October 1974, President Gerald Ford signed the Energy Reorganization Act. The Atomic Energy Commission was split asunder like a fissioning uranium atom: The production and energy research and development aspects became the province of a place-holder agency called the Energy Research and Development Administration, later transmogrifying into the U.S. Department of Energy. The puny and often conflicted regulatory function of the AEC, which licensed new power reactors, became the U.S. Nuclear Regulatory Commission.

The Joint Committee on Atomic Energy died a lingering death alongside the AEC. Shortly after Ford signed the reorganization law, the House adopted a proposal from Richard Bolling of Missouri, a senior Democrat on the House Rules Committee, which began chipping away at the joint committee's expansive jurisdiction. The Bolling Amendment gave authority for non-military aspects of nuclear power to the House Interior Committee. In August 1977, both the House and the Senate formally abolished the joint committee, following a scathing report the year before by the reformist group Common Cause, titled "Stacking the Deck," and outlining how the committee had become a "huckster for the nuclear power industry."

4. What Friendly Atom?

By 1952, when the Republicans under Eisenhower recaptured the White House, the House of Representatives, and the Senate, civilian control of nuclear energy—at least on paper—was well established. The AEC was up and, if not running, then at least lurching ahead.

David Lilienthal had stepped down as chairman of the commission in February 1950 and was succeeded by Gordon Dean, whom Truman appointed to the commission in 1949 to fill one of the two vacancies created when commissioners Robert Backer and William Waymack resigned. Dean, a forty-six-year-old lawyer, had only one pertinent qualification. He had been a law partner of Sen. Brien "Mr. Atom" McMahon, the Connecticut Democrat who was a key player in the passage of the 1946 Atomic Energy Act (the bill was known as The McMahon Act). Truman designated Dean chairman when Lilienthal left.

Dean's tenure, which lasted past the Eisenhower triumph of 1952 and ended on June 30, 1953 (the end of the federal fiscal year 1952), was largely unremarkable, except for some anger he elicited from the military. *Time* magazine reported in January 1952, "The most violent criticism of Dean and his colleagues comes from the military. None of the brassier brass hats wants to be quoted, but out of the Pentagon's propaganda orifices comes a continuous stream of bitter unofficial protest. It is not against the AEC's scientists. These the Pentagon loves: they think up atom bombs. It is against the AEC itself, a snooty civilian agency that speaks a language which few generals understand, whose power is enormous, and whose lowliest employees are not afraid to talk back to the Pentagon."

The military was pushing to expand the U.S. arsenal with bigger and more deadly weapons, while the AEC was recommending smaller bombs with more bang-per-pound and -dollar, including battlefield tactical nukes.

The magazine cited Gen. Curtis Lemay, head of the Air Force Strategic Air Command, which was flying the enormous B-36 bombers, as scornful of smaller bombs. Army General Joe Collins, an artillery man, was also skeptical of small, tactical nuclear weapons, preferring 150-ton shells that could be fired from enormous Army artillery barrels.

The irony of this contretemps highlighted in *Time* magazine was that the military was getting virtually everything it wanted from the AEC, and would continue to enjoy its accustomed deference and largess. What was lacking was any progress on civilian power.

As the Eisenhower administration moved into Washington, the promises of the early days of nuclear madness had gone largely unfulfilled. There were no houses heated by small nuclear reactors, no cars or locomotives with atomic engines. And there never would be such imaginative, or hallucinatory, devices. At that point, a nuclear submarine was still considered visionary, and many experts, including some in the Navy, didn't think it was possible.

Moreover, the most practical and obvious potential civilian fruit of nuclear energy—electricity—was nowhere to be seen The civilian power industry had shown complete indifference to investing in nuclear power , which promised to be expensive and difficult to build. To an industry accustomed to cheaply burning coal, the transition seemed to impose unnecessary burdens. Without a compelling business case that showed clear financial incentives to building electricity-generating nuclear plants, no one in the industry dared to make the leap.

Eisenhower, whose military credentials were above challenge, felt strongly that the nation had to refocus its atomic attention from weapons to defend the nation to technology to support the country's welfare. Ike didn't want to abandon military nuclear power, but he wanted to deliver on the civilian promises of the atom. His administration moved quickly to assert control over the AEC.

The White House named Strauss, a true-blue, pro-military and -business conservative, to lead that transformation. Strauss had been feuding with Lilienthal for months in 1949 and 1950, largely over secrecy and security. Strauss felt Lilienthal was too willing to share the secrets of the atom with U.S. allies, such as Great Britain; he also pointed to what he perceived as Lilienthal's lack of concern about personnel security policies dating back to the Manhattan Project, which didn't do enough to weed out former Communists from the enterprise.

The final dispute between Lilienthal and Strauss was over development of the hydrogen fusion bomb, also known as the Super. Lilienthal and the scientists on the General Advisory Committee—led by Oppenheimer—opposed work on the weapon, fearing it would result in an even more ferocious arms race with the Soviet Union than was already underway.

Strauss, with backing from physicists Ernest O. Lawrence and Edward Teller, supported development of the bomb, with its potential to be at least one thousand times more powerful than the bombs that brought Japan to its knees. With Gordon Dean's support, Strauss persuaded Truman to support development of the H-bomb. Strauss then resigned from the AEC, leaving only months after Lilienthal's final day. With White House backing, the AEC directed a successful program to create the most terrifying and dangerous weapon the world had ever seen. The Soviets, convinced that the purpose of the weapon was to annihilate its communist regime (Teller would probably not have disagreed with that aim), mounted a successful crash program to develop its own thermonuclear bomb—and the arms race was on.

Strauss was a conservative Republican who backed Ohio Sen. Bob Taft for the 1952 GOP nomination that Eisenhower won. After Eisenhower's victory, he began reaching out to Taft Republicans. Iowa's Hickenlooper, a key player on the Joint Committee on Atomic Energy, suggested Strauss as AEC head. Taft pushed Strauss for a top Pentagon job, and former president Herbert Hoover suggested his former private secretary would be a dandy commerce secretary.

Events moved quickly. The victorious former general nominated Strauss, a former wartime admiral, to chair the AEC on June 24, 1953; the Senate confirmed the nomination three days later. On July 2, Strauss took the oath of office.

On December 8, 1953, Eisenhower addressed the United Nations General Assembly in New York, where he gave his famous Atoms for Peace speech. The oration was directed as much toward the American people as to the rest of the world. Citing his military experience, Eisenhower said he felt the need to unburden himself "by saying to you some of the things that have been on the minds and hearts of my legislative and executive associates, and on mine, for a great many months: thoughts I had originally planned to say primarily to the American people."

Pondering the abyss of nuclear annihilation, Eisenhower proposed an international body to harness and control the power of fissionable elements, turning them into instruments of peace. "So my country's purpose is to help us move out of the dark chamber of horrors into the light," Eisenhower said, "to find a way by which the minds of men, the hopes of men, the souls of men everywhere, can move forward towards peace and happiness and well-being." The United States, Eisenhower said, "would seek more than the mere reduction or elimination of atomic materials for military purposes. It is not enough to take this weapon out of the hands of the soldiers. It must be put into the hands of those who will know how to strip its military casing and adapt it to the arts of peace.

"The United States knows that if the fearful trend of atomic military build-up can be reversed, this greatest of destructive forces can be developed into a great boon, for the benefit of all mankind. The United States knows that peaceful power from atomic energy is no dream of the future. The capability, already proved, is here today. Who can doubt that, if the entire body of the world's scientists and engineers had adequate amounts of fissionable material with which to test and develop their ideas, this capability would rapidly be transformed into universal, efficient, and economic usage?"

There was no reason to suspect that Strauss was the wrong choice to implement Eisenhower's vision. He was a forceful, successful executive with a background that suggested he could transform the AEC into a force for peace as well as the forge of the weapons of war.

Strauss was a fabulously successful investment banker who started out in life selling shoes wholesale for his Richmond, Virginia, family business. A good student interested in science in high school, he had hoped to enroll at the University of Virginia to study physics, but a recession in 1913–1914 forced him onto the road selling shoes, where he was quite successful. His diligence in flogging footwear saved the family business.

In 1917, the supremely confident Strauss traveled from Richmond to Washington, DC, to apply for a job as private secretary to Herbert Hoover, whom President Woodrow Wilson had asked to take over the task of feeding American troops in World War I. The brash young man from Richmond got the job, impressing Hoover with his hustle. At the end of the war, Hoover—with Strauss in train—organized relief for war-torn Europe.

In 1919, Strauss joined the investment banking firm of Kuhn, Loeb in New York, where he made a considerable fortune and became active in Republican politics as a Hoover devotee. When a second worldwide war broke out in 1942, Strauss, a long-time Navy reserve officer, volunteered for service and eventually earned the rank of admiral, as an expert in logistics.

Strauss's successes in business and the military, and his Republican connections, led Truman in 1946, after the passage of the Atomic Energy Act, to name the Virginian to the Atomic Energy Commission. Strauss, the wannabe physicist, possessed an optimistic, even romantic, view of the role of the atom in the United States' future. After he resigned from the AEC in 1950, scuttlebutt immediately had Strauss becoming chairman should the Republicans claim the White House in 1952. Strauss did nothing to chill the rumor mill. When Eisenhower named him to head the commission, there was no reason to doubt the ability of Strauss, who had demonstrated skill and loyalty at every level of business and government service.

Strauss, the accomplished businessman, had very different ideas than David Lilienthal about how to chair the five-member commission and how to run the sprawling, still infant agency. Strauss, accustomed to hierarchy and top-down decision making, had been uncomfortable with the way Lilienthal tried to run the AEC commission as a circle of equals. Strauss biographer Richard Pfau summed up the new approach to running the AEC as follows: "Strauss brought a new style of management to the Atomic Energy Commission. Accustomed to action rather than debate, he replaced the informal, rambling, seminars of his predecessors Lilienthal and Dean with crisp, formal sessions designed to produce quick decisions rather than slowly evolving consensus. Forceful, determined, and impatient once he had made up his own mind, Strauss dragged the other commissioners along with him."

While Eisenhower and Strauss hoped to redirect the AEC toward more civilian uses of atomic energy, that never came to pass during their tenure in Washington. Rather, military applications continued to drive the attention of the commission and the administration and soak up the endless amounts of money provided by the joint committee. The AEC prioritized building and diversifying the atomic weapons stockpile, developing new weapons including nuclear submarines and nuclear bombers, and rooting out the alleged Communist menace within the agency.

Strauss entered into this set of priorities led both by forces beyond his control and by firm intention. He was convinced that Groves had permitted lax security during the Manhattan Project, in part because it was necessary to attract the best scientists to the endeavor. Like many of the intelligentsia of the time, some atomic physicists and chemists had been sympathetic to the left. In some cases, they had actually been Communists (although none of the key American scientists who worked on the bomb were in any way associated with spying for the Soviet Union[20]).

Strauss participated in the hunt for Red scientists with gusto. He had a personal grudge to bear against J. Robert Oppenheimer, who had belittled Strauss in public over the issue of developing the hydrogen bomb. Oppenheimer opposed the weapon and considered Strauss a fool for pushing it. Oppenheimer's arrogant scorn for those whose views he believed were wrong was legendary. Strauss, with Teller whispering in his ear, was convinced that Oppenheimer was a dangerous leftist who should be purged from any access to the holy secrets of the atomic weapons program. A 1947 security report to Groves had pointed to Oppenheimer's ties to leftists (his wife) and Communists (his brother), but Groves concluded that Oppenheimer was a loyal American—and crucial to the development of an atomic weapon.

Strauss in 1954 initiated security proceedings against Oppenheimer, a controversy that dominated his time as head of the AEC. While ultimately successful in forcing Oppenheimer out of the AEC's General Advisory Committee[21], the event tainted Strauss from that time forward and soured the public image of the commission. The vendetta against Oppenheimer plunged the AEC for the first time into partisan politics. Strauss biographer

---

20 The only top-notch scientist found to have spied for the USSR was Klaus Fuchs, who provided the Soviets with information on Teller's Super project. He was a citizen of Great Britain and a committed Communist. Whether the information he provided to his puppet masters helped the Russians develop their H-bomb is a matter of considerable controversy, as the design he stole was Teller's first, flawed attempt to initiate the fusion reaction.

21 Strauss had the shameful assistance of Edward Teller, who helped immeasurably to throw his scientific rival under the national security bus.

Pfau wrote, "Until the summer of 1954 the Atomic Energy Commission conducted its affairs free of partisan politics."

The Oppenheimer security hearings were a travesty of innuendo, personal animus, and inquisition. The attack on Oppenheimer included wiretaps, clandestine surveillance, and routine violations of the attorney-client privilege. At a time when the Soviet Union had a tarnished reputation for show trials of political opponents of its murderously repressive regime, Strauss gave critics a milder, but no less telling, analog. His biographer, who was mostly sympathetic to Strauss, cut to the bone on his assessment of the Oppenheimer vendetta: "Determined to eliminate Oppenheimer, Strauss spared nothing. He failed to take account of the long-range effect of his action. For Strauss, no sacrifice was too great, even the sacrifice of himself. In the end, the Oppenheimer case cost the United States the services not only of Oppenheimer but also of Strauss."

Strauss also seemed to go out of his way to court controversy. He quickly enmeshed the Eisenhower administration in a nasty battle with advocates of publicly-owned electric utilities over who should build new generating plants to supply electricity to the AEC's burgeoning Oak Ridge weapons lab. The dispute turned into the Eisenhower administration's first, and perhaps most brutal, partisan conflict, with Strauss fanning the flames for the Republicans. It became known as the "Dixon–Yates" controversy.

On December 2, 1953—six days before Eisenhower's Atoms for Peace speech—Strauss met with the president's budget chief, Joseph M. Dodge, who informed the AEC chief that the TVA needed more electricity in order to continue to serve its territory, particularly the city of Memphis. The AEC's sprawling Oak Ridge weapons laboratory and the AEC uranium enrichment plants there and at Paducah, Kentucky, were the TVA's largest industrial customers. The bulk of the TVA's electric power went to publicly-owned distribution utilities—electric departments owned by cities and towns in the region and rural electric cooperatives. Oak Ridge consumed about a third of the power the TVA generated from its hydropower dams and coal-fired power plants.[22]

---

22 During the 1960s and 1970s, the TVA would become the most enthusiastic utility in the country for nuclear plants, ultimately ordering a total of seventeen units from virtually

Dodge explained to Strauss that the administration really didn't like the idea of allocating federal funds to build power plants that private industry could build. The TVA exemplified what Republican orthodoxy defined as "creeping socialism," and on this matter Ike was ultra-orthodox. Strauss pledged to Dodge that the AEC would find a way to buy its power in the TVA region from the private sector, as it did elsewhere in the United States.

Strauss clearly didn't know what he was getting into with this seemingly uncontroversial decision. What followed was a reprise of a battle that Lilienthal had waged during his years as the head of the TVA. In the earlier rumble, the TVA had prevailed over the private utilities and their champion, Wendell Wilkie, the failed GOP presidential nominee in 1940, putting the Tennessee Electric Power Co. out of business.

It was no surprise that private power companies were quick to come up with a plan to shut the TVA out of building new power generating plants to supply the AEC in the middle south. Strauss hooked up with an old acquaintance from his investment banking days, Adolph Wenzell of First Boston Corp. Strauss didn't know that Wenzell was working both for the giant investor-owned Middle South Utilities and as a consultant to the White House Bureau of the Budget, a clear conflict of interest that, when later revealed, made Strauss look venal.

Middle South's Edgar Dixon and Eugene Yates of The Southern Co., a large Atlanta utility holding company, proposed a contract with the AEC whereby a new company—Mississippi Valley Generating Co.—jointly owned by Middle South and Southern, would build generating capacity and sell the power to TVA for the city of Memphis, freeing up power to flow to the AEC. Strauss signed the contract setting off a fight with the TVA, its powerful Washington allies in the American Public Power Association, the National Rural Electric Cooperatives Association, and important congressional Democrats. Among the outraged Democrats were Sen. Clinton Anderson of New Mexico, long a force on the Joint Committee on Atomic Energy, and politically ambitious Estes Kefauver of Tennessee.

---

every major purveyor of nuclear generating technology, including Westinghouse, General Electric, Combustion Engineering, Babcock & Wilcox, and General Atomic.

Democrats had regained control of Congress in the 1954 off-year elections and Anderson became JCAE chairman.[23] In 1955, acrimonious hearings on Dixon–Yates in the Senate Judiciary Committee's antitrust subcommittee, with Kefauver as chairman, brought Wenzell's conflicts to light, damaging Strauss in the process. Predictably, the White House shortly announced that it would cancel the contract, but only after considerable damage to Strauss and the Eisenhower administration.

Strauss's final controversy came over continued testing of nuclear weapons and the fears of nuclear fallout . This issue simmered as the United States and the Russians scaled up above-ground tests during the early 1950s. In the fall of 1954, it boiled over after U.S. H-bomb testing dropped radioactive debris on the Japanese tuna fishing boat *Daigo Fukuryū Maru* (Lucky Dragon) and South Pacific islanders. Throughout the controversy, and later when defending the need for continuing atmospheric testing, Strauss displayed his rigid and sanctimonious nature.

On March 1, 1954, the AEC detonated the first lithium deuterium-based (dry) H-bomb[24], known to the bomb mavens of Los Alamos as Shrimp, for its comparative size. The test, known as Castle Bravo, took place on Bikini Atoll in the Marshall Islands. It was a tremendous technical success, as the weapon released an explosion calculated at 15 megatons, three times the yield that AEC scientists had estimated.

The bomb was so powerful that it trapped observers in bunkers well outside the projected range and rained radiation on U.S. Navy ships observing the test from sea. Harold Brown, an AEC atomic bomb expert and later secretary of defense in the Carter administration, said that "Shrimp went like gangbusters." AEC bomb builder Marshall Rosenbluth told historian Richard Rhodes, "I was on a ship that was thirty miles away and we had this horrible white stuff raining out on us. I got ten rads[25] of radiation from it. It was pretty frightening."

---

23 The Democrats continued to control the House for the next forty years and lost control in the Senate only briefly, following the 1980 election.

24 Early H-bombs used liquid deuterium (a hydrogen isotope) in order to simplify the research, but the bomb-makers were planning to switch to the dry design as a second-generation thermonuclear weapon.

25 The equivalent of about 10 chest X-rays.

The event didn't particularly frighten the twenty-three-man crew of the unlucky Japanese tuna boat, which had been working some sixty miles east and downwind of Castle Bravo. They were unknowingly drenched with radioactive ash and dust from the blast for three hours. The men scooped up the detritus as it fell on their vessel and bagged it with their bare hands. When the ship pulled into its home harbor in Yaizu, Japan, on March 14, all of the crew showed symptoms of radiation poisoning. They were nauseous, with bleeding gums, suffering from headaches and eye pain. One of the crew members eventually died, apparently of a secondary infection that is common with radiation poisoning.

Strauss's performance in the ensuing controversy was disgraceful. When the event surfaced in Japan, it created an understandable uproar. The United States was again, even if inadvertently this time, attacking Japan with nuclear weapons. In Washington, Strauss washed his hands of any U.S. responsibility. The test never got out of control and the Japanese ship "must have been well within the danger area," he said. When the crewman died in September, Strauss insisted that the death was a result of "hepatitis from antiquated medical techniques." He told Eisenhower press secretary James Hagerty that the Japanese vessel was "a Red spy ship." There was not a shred of evidence to support any of Strauss's assertions.

Called before the JCAE in 1955, as scientists began questioning the value of the atmospheric tests in light of the increasing concern about radiation, Strauss was adamant about the need to continue testing. "The degree of risk," he said in a press release, "must be balanced against the great importance of the test program to the security of the nation and the free world." But the AEC consistently downplayed the risk, as nuclear physicist Ralph Lapp demonstrated convincingly in an article in *Science* magazine.[26]

In a 1955 article in the *Bulletin of the Atomic Scientists*, Lapp openly discussed the impact of radioactive fallout from the hydrogen bomb. Hickenlooper, the Iowa Republican and stalwart on the joint committee, was so disturbed by Lapp's findings that he instructed the AEC to investigate Lapp for revealing classified information. Strauss complied, although Lapp was eventually cleared of any charges.

---

26 Lapp later wrote an excellent book on the Castle Bravo test, *Twenty-three Fishermen and a Bomb: The Voyage of the Lucky Dragon*, Harper & Bros., 1958.

During both the Oppenheimer show trial and the Dixon–Yates controversy, Strauss frequently let his prickly personality and inherent arrogance govern his behavior. That succeeded in making a lasting enemy of New Mexico's Anderson, who had no shortage of arrogance and bile himself. They clashed repeatedly during the testing hearings.

Strauss left the AEC in 1958 and became a personal aide to Eisenhower on national security issues. In 1959, Eisenhower nominated him to be secretary of commerce. Strauss's Democratic enemies—particularly Anderson—seized on the nomination as a chance to gain retribution and bloody the Republican administration in the process. Anderson turned all his skill and considerable vitriol against Strauss and defeated the nomination.

Reporting on the fight over the Commerce Department nomination, *Time* magazine aptly summarized Strauss's tenure on the Atomic Energy Commission:

> Strauss's five years as chairman of the Atomic Energy Commission resounded with an endless rumble of controversy. The wounding wrangle that followed the suspension of physicist Oppenheimer's security clearance made Lewis Strauss an unforgiving enemy among the nation's scientists...He drew much of the blame for AEC's heavily attacked (and long since cancelled) Dixon—Yates contract, under which a private utility was supposed to build a power plant at West Memphis, Ark., right in the jealously guarded public-power domain of the Tennessee Valley Authority. He outraged stop-the-tests advocates by urging continued nuclear tests, with emphasis on developing "clean" weapons.

5. The Atomic Chimera

Throughout the Strauss controversies, the nation's leaders had embarked on restructuring the Atomic Energy Commission to reflect Eisenhower's peaceful aims and in response to a precipitous slide in public support for atomic energy. The government had not been able to deliver on its grandiose postwar promises for atomic power, and the public's patience was growing short.

In 1953, the University of Michigan's renowned Survey Research Center looked at public attitudes toward atomic energy, publishing their results in the January 1, 1954 issue of *Science* magazine. The study found broad pessimism about the prospects for non-military uses of nuclear power. The researchers said the polling indicated that "it does not appear that the development of atomic energy and the allegedly revolutionary social and economic changes which it portends have captured the imagination or stimulated the curiosity of the majority of the law population."

The conclusion, perhaps expected but hardly persuasive, was that the "lay population" was "ill-informed about atomic energy apart from the atomic bomb." In a sophisticated variant of "blame the victim," the analysis concluded condescendingly that those who had a negative view of the non-military fruits of the atom were "chronic pessimists" and characterized by insecurity and low in self-confidence.

More to the point, the *Science* article also noted in passing that "the development of atomic energy, up to this point, has been tied largely to wartime needs and military demands. Peaceful applications have been comparatively modest and exceedingly specialized. Their consequences have not directly entered the lift of the average citizen. His automobile is not run by atomic power; atomic power has not changed his job, his diet, his house, or his recreation."

The Eisenhower administration, the joint committee, and the AEC were aware of a disconnect between promise and delivery. Their diagnosis focused, in part, on the shortcomings in the 1946 Atomic Energy Act, which kept a tight government monopoly over all things atomic. Under the act, it was a crime for public citizens to possess nuclear materials. All research and development, for both military and civilian uses of atomic energy, was in tightly-controlled government hands at the AEC. No private citizens could hold patents involving use of fissionable material. The historians Mazuzan and Walker noted, "Civilian direction of the agency did not mean liberalized control of the atom by the government." Any research and development outside the government required AEC approval. The AEC had special authority to classify information that extended far beyond that employed by the military. The atomic agency could criminalize possession of what it alone defined as "restricted data." That gave the agency the authority to classify information that had already appeared in the public record, a practice that was common inside the AEC. While this gave the AEC, and its overseers, tremendous power, it also meant limiting the range of ideas and breadth of experience that could be brought to new ways of using the power of fission and fusion.

Soon the Eisenhower administration began promoting reform of the 1946 act; Congress quickly agreed to consider changes in the law governing atomic energy. In February 1954, following his U.N. speech two months earlier, Eisenhower sent a message to Congress calling for revising the 1946 law to make "it possible for American atomic energy development, public and private, to play a full and effective part in leading mankind into a new era of progress and peace." On August 30, Eisenhower signed the 1954 Atomic Energy Act, a major revision of the 1946 law, opening up the nuclear vault to greater civilian access, hoping to promote peaceful uses of the enormous power of the atom.

But old patterns and habits refused to change. While their lips said peace, the hearts and minds of the nuclear power establishment continued to say, and mean, war. This tension remained throughout the 1950s and 1960s and well into the 1970s.

Overall, Strauss's cockeyed optimism flawed his vision of the future. That's no surprise. The nuclear delusions were widespread, and the ability of government in all branches to foresee the future has a wretched record, as do beard-stroking pundits and denizens of learned think tanks. Princeton psychologist Philip Tetlock, starting in the 1980s, began studying 284 men and women who make their living predicting the future and scoping out political and economic trends. He published the results in his 2005 book *Expert Political Judgment*. These were folks steeped in their fields, gurus and mavens all. They proved unable to make consistently valid predictions. As Danish physicist Niels Bohr said, "Prediction is very difficult, particularly about the future."[27]

Some of the atomic vision made sense. Nuclear power was well-suited for Navy applications, where weight was not a problem and cost was no barrier. Use of isotopes in medicine was a fine application, giving physicians the ability to trace the activity of drugs in the human system and see deep into body tissue. Even reactors to produce commercial electricity had considerable value, though it never approached being "too cheap to meter."

But much of the enthusiasm about life with the tamed atom made no sense. Frequently the most harebrained schemes for atomic energy ignored the very real properties of radioactive materials: ionizing radiation can sicken and kill. Take the notion of atomic-powered cars, which was simply radioactive moonshine. Imagine a 50-car pileup on an interstate in the California fog if the cars had glowing, radioactive power plants? The results are too gruesome to detail. Detroit's flirtation with nuclear-powered automobiles never moved beyond passing notes and giggling.

The intriguing idea of small nukes that could heat homes and provide onsite electricity also proved to be an illusion. The Army pursued this with its SL1 program at the Idaho testing site. The Army wanted a small reactor that could serve remote bases in places like Alaska with their entire needs for heat, water, and electricity. The result was the worst nuclear disaster in U.S. history. The test reactor in Idaho, for still-unknown reasons, lost control and killed three soldiers in a grisly catastrophe. One soldier was impaled through the groin and pinned to the top of the reactor building by an errant control rod. The accident got little public attention. But it marred the Atomic Energy Commission's reputation for many years.

---

27 This quote has also been attributed to former New York Yankees catcher and shade tree philosopher Yogi Berra, although this is likely apocryphal.

The pundits and politicians largely had more big-muscled roles in mind for nuclear power than how to heat Alaskan military outposts. They pursued a grander agenda. Bombers powered by nuclear reactors would command the air, circling at high altitudes, patrolling enemy skies with the ability to rain death and destruction at will. Nuclear explosives would move mountains, dig channels, carve out new harbors, and liberate vast untapped reserves of oil and natural gas. Nuclear power plants would make steam, generating vast amounts of electricity, while generating their own fuel, producing more fissionable material than they would consume. Ultimately, the power of the H-bomb would turn into endless, clean, free electricity.

Radioactivity? No problem. Scientists promised they could keep the radioactive emissions from nuclear energy at bay. Gordon Dunning, the AEC's in-house expert on radiation, was always sanguine in public about the threat posed by exposure to radiation through nuclear technology and, most particularly, from fallout resulting from nuclear weapons tests. Writing in *Scientific American* in 1955, Dunning assured the public that atmospheric weapons testing posed no threat to the public. The dreaded word "cancer" never appeared.

Strauss summoned the vision of the friendly atom, which would transform American society entirely for the better, in his 1954 speech to science writers in New York. That optimism echoed his boss, President Eisenhower, who had asserted to the United Nations the previous year :

The United States knows that if the fearful trend of atomic military build-up can be reversed, this greatest of destructive forces can be developed into a great boon, for the benefit of all mankind. The United States knows that peaceful power from atomic energy is no dream of the future. The capability, already proved, is here today. Who can doubt that, if the entire body of the world's scientists and engineers had adequate amounts of fissionable material with which to test and develop their ideas, this capability would rapidly be transformed into universal, efficient, and economic usage?

Meanwhile, out in the desert reaches of the western United States, Atomic Energy Commission scientists and engineers were being increasingly supplied with fissionable materials, with full-scale government backing. They were being tasked with turning the power of the bomb into several forms of aviation, for both military and peaceful uses. The results belied the technological optimism of the post-Manhattan Project age.

46

Up in the Air: Flights of Radioactive Fancy

In the first book of the Tom Swift Jr. series, *Tom Swift and His Flying Lab*, young Tom designed and built a gigantic, atom-powered flying laboratory, which included a full-scale nuclear and chemical lab, sleeping quarters for a large crew, and a gourmet galley. The multi-floored aircraft also utilized jet technology for vertical takeoff and landing and could house and launch a small, jet-powered craft as well as a helicopter.

In building the nuclear flying laboratory, it took Tom about three months from conception to flight. Dubbed the *Sky Queen*,[28] Tom's nuclear craft could fly for months at 1,000 mph, at altitudes just short of outer space.

Understandably, Victor Appleton chose to start his postwar literary venture with flight. Flying has always fascinated Earth-bound man; children commonly dream that they can fly. In only fifty years, airplanes had grown from fantasy to flying flivvers to futuristic weapons systems that arguably delivered the war into American hands. Nuclear flight seemed to be the next logical step and perhaps the stepping off point for mankind to venture into space.

Unfortunately, while producing large amounts of radiation and consuming even larger amounts of taxpayer dollars, the development of nuclear flight, both atmospheric and stratospheric, remained firmly anchored to the ground.

---

28 No doubt a reference to the popular Sky King radio show of the day.

6. The Bomber to Nowhere

An enormous building squats on the high desert of eastern Idaho, the most tangible artifact of one of the more daft and feckless federal research and development programs of the 20th Century. Built in 1959 for $8 million, the building looks like a giant oil barrel half-buried lengthwise in the sand. It spans a clear space of 320 feet by 240 feet, or two football fields side-by-side.

Tom Swift's vision, scaled back to reflect some reality, was just what the Army Air Corps had in mind in 1946 when it launched the most ambitious and foolish nuclear project in U.S. military history, the nuclear-powered bomber. While Tom's nuclear airplane was pure fiction, the Air Force A-plane was all too real. Tom won the race—his airship flew. U.S. taxpayers lost in the real world beyond boys' books.

An enormous building on the high desert of eastern Idaho was designed to be the hangar for the nation's nuclear-powered bomber, a government project that spanned fifteen years, cost at least $1 billion[29], and produced just that: a giant hangar in Idaho. Today, its ironic mission is to house depleted uranium, used in military munitions, particularly tank-penetrating artillery shells.

Had a runway been built for the atomic airplane, it would have run some twenty-three thousand feet (more than four miles) in order to get the behemoth off the ground. Whether the plane would have been able to land at the strip is unknown, and probably unlikely. But because of the risk of radiation from a crash, it would never have been allowed to land—on land.

Years of toil and countless dollars were spent attempting to achieve what was neither technologically achievable nor militarily useful. Clearly,

---

29 That's about 20 billion in today's dollars.

from the start of the project, the Air Force simply didn't understand much about nuclear physics. A version of Tom Swift's pie-in-the-sky flying laboratory had emerged from largely impractical and hopelessly optimistic government and military fantasy, all the while providing plenty of laughs and much head-scratching along the way. It's also a tale largely forgotten in the mists of nuclear history, which focuses on the successes of the government-industry nuclear partnership following World War II.

It was clear from the start of the nuke airplane project that the Air Force simply didn't understand much about nuclear physics, any more than did Tom Swift. Tom can be forgiven; he was an 18-year-old with no education in physics, a fictional creation of men of imagination with no known scientific or technical skills. The U.S. government cannot be forgiven. It hired the best physicists in the world. They were real.

A joint venture of the United States Air Force (originally the Army Air Corps); the National Advisory Committee on Aeronautics (predecessor to today's NASA); and the Atomic Energy Commission—it classically illustrates postwar technological optimism, as policymakers in Congress and the executive branch viewed the project through rose-colored glasses for a decade and a half.

It's also a classic case of fierce inter-service rivalries; conventional congressional pork-barrel politics tweaked to a high degree; and nasty turf warfare among various executive and legislative arms of the federal government. The nuke bomber offers a model of how modern, earmarked politics can produce entirely irrational outcomes, such as the move by the Alaska congressional delegation in 2005 to build a "bridge to nowhere."

Following World War II, Congress, the Air Force, and the AEC attempted—against sound scientific advice—to build a bomber to nowhere, and failed entirely. Science trumped politics, much to the dismay of the politicians, who moaned long after their failure that they weren't to blame and that their vision wasn't blurred by fantasy. The enthusiasts insisted, in the face of mounting evidence, that nuclear flight was just around the corner—if the taxpayers just had the will to stay on the runway.

The attempt to build an atomic bomber began as a radioactive dream among Air Force strategists and some nuclear energy experts not long after the United States dropped atomic bombs on Japan in 1945. The impetus for both the Navy's successful nuclear submarine program and the Air Force bomber-to-nowhere took off at the same time President Truman signed the Atomic Energy Act. This meant the services would have to work with the AEC, a civilian agency, to develop their atom-powered engines. That proved to be an easy obstacle to overcome, as the AEC was quick to endorse and support the military's plans for use of nuclear power beyond blowing up cities and civilians.

The Navy promptly assigned an up-and-comer, the prickly-but-productive Capt. Hyman Rickover, to run its submarine program. Submarines made sense for the application of nuclear energy. Rickover was a hard taskmaster, a fine judge of talent, and a most practical engineer. He was not essentially a military man, but a man who would turn his engineering and organizing talents to a military end.

The Air Force was dominated by men who grew up dropping bombs and were military to their bones. Both Curtis Le May, the cigar-chomping, hard-charging head of the Strategic Air Command, who earned his chops as a bomber pilot, and Air Force chief Hap Arnold were strategic bombing enthusiasts. The Air Force also had a reputation of being less technically adept than its rivals in the Army and Navy. Military historian Stephen Budiansky, for example, wrote that Arnold "had often frustrated his more technically competent subordinates through his ignorance of basic science and mathematics (he didn't know how to use a slide rule, one recalled, and would become manifestly impatient with technical explanations), but by the Second World War he had acquired a sort of layman's admiration for scientific wizardry."

Le May's atomic aircraft, in his mind, would become the ultimate force in strategic bombing. The Air Force brass clung to the notion that strategic bombing would win the wars of the future, despite compelling evidence to the contrary. The vision of the nuclear bomber that beguiled him and the other proponents was a craft that could fly at supersonic speeds, staying aloft for weeks or months at a time without the need to refuel. That amount of fuel that conventional propeller-driven bombers carried slowed them down. High-speed jets burned fuel so fast they couldn't stay up long. Flying nukes seemed to solve both problems.

Public notice of the atomic bomber program came in early 1947. A *New York Times* article in mid-February mentioned a ten-day, "non-publicized" (meaning secret) conference on nuclear aircraft at Oak Ridge that ended

February 13. Oak Ridge was a center for reactor development for the AEC, dating back to the Manhattan Project. Insiders referred to the meeting as the "college of nuclear knowledge." The *Times* article said, "Scores of high-ranking Army officers, top-flight nuclear scientists, and representatives of aircraft engine and airframe manufacturers and chemical companies were at the meeting." The program they were creating, said the piece, would be known as NEPA, for Nuclear Energy for Propulsion of Aircraft.

The Air Force and AEC's confidence that they could accomplish this was stunning, given the daunting technical obstacles of the project. In October 1948, David Poole, an Oak Ridge engineer, told the Baltimore Society of Automotive Engineers that the "theory of an atom-driven airplane was 99 percent perfected. The time has come when we can no longer afford not to have atomic aircraft." He was laughably off base.

The Air Force brass thoroughly embraced the leap of nuclear faith. In 1947, they were predicting it would only take five years to turn what were essentially paper nuclear-powered airplanes into demonstrated flight. Their motto was Fly Early.

There were notable skeptics, even in the early halcyon days of nuclear non-flight. Ironically, the first article printed on the topic in the *New York Times*, January 26, 1947, quoted noted radiation expert Ralph Lapp, then at the War Department,[30] as saying that while a nuclear airplane was theoretically possible, the most difficult practical task would be shielding the crew against radioactivity. Lapp's doubts were on the money.

Some of the people who best understood atomic power, the bomb builders, were very skeptical of the nuclear bomber. Oppenheimer, as chairman of the AEC's General Advisory Committee, repeatedly ridiculed the plane, dismissing it as "hogwash." The *New York Times* in 1953 quoted Edward Teller, the father of the hydrogen bomb and an enthusiast of most things nuclear, expressing scorn for the atomic bird. With a thick eastern European accent, Teller observed that the plane "must not crash. A nuclear-powered airplane must not be flown near heavily populated areas." Teller added sardonically, "But this is only assuming there will be nuclear-powered airplanes." Those who knew Teller over the years can imagine his impressive eyebrows lifting in scornful emphasis.

---

30 Today's Defense Department.

In 1948, not quite a month after Oak Ridge's Poole gave his optimistic views to the Baltimore engineers, Hanson Baldwin, the *Times* fine military affairs writer, poured very cold water on the Air Force's hot atomic airplane fever. Baldwin wrote a long and perceptive article, referring to Poole's Baltimore appearance. "This enthusiastic—but highly inaccurate and misleading—account in no way represents the actual progress made by NEPA or the present prospects," wrote Baldwin.

"The enthusiasm is well justified; there will be atomic-powered planes someday, but not soon. Atomic powered planes are not possible, regardless of the effort expended on them, within the near future, certainly not within the next three years, probably not for at least a decade. Not all the problems are fully known, much less answered. Stationary land power plants and atomic power plants for naval vessels are under study and these offer hope of considerably quicker development." Even Baldwin's pessimistic prognostication proved wildly optimistic.

With the Air Force, the AEC, and Congress backing urgent development of the atomic airplane, and high-profile critics such as Oppenheimer doubting its feasibility, a group of forty experts assembled at the Massachusetts Institute of Technology in 1948 to examine the issues, led by physicist Walter G. Whitman. Calling themselves the Lexington Project, they concluded that the plane would take at least fifteen years and a billion dollars to develop, by which time missile technology might make the atomic bomber unneeded. These experts pointed to a series of difficult technical obstacles the plane would have to overcome, including weight, shielding, the need for novel materials, and the like. Clearly, the group thought the atomic plane was a bad idea.

Nevertheless, the Air Force and the AEC pushed ahead with typical hubris and enthusiasm, claiming the Lexington Project had validated their vision of the nuclear bomber. Congressional support and funding followed, prioritizing the NEPA program. The Joint Committee on Atomic Energy oversaw the project.

The Oak Ridge National Laboratory in Tennessee was working on aircraft reactor designs. By then known as the Aircraft Nuclear Propulsion (ANP) project, the Air Force was contracting with air frame companies and reactor vendors in politically-important Congressional districts around the county. They appeared to make their contract decisions in such a way as to generate maximum support for their project. This, of course, was no surprise and characteristic of the way the AEC often developed its major projects.

General Electric built a large plant in Evendale, Ohio, near Cincinnati, to work on the reactor engineering. In mid-1952, the AEC announced it would spend $33 million at the National Reactor Testing Station in eastern Idaho, where the plane would be housed and tested. Other major contracts went to California and Connecticut.

The Truman administration, working with the joint committee and a Democratic Congress, gave strong support to the Air Force project. Truman's second defense secretary, Louis Johnson, was besotted with air power, buying in completely to LeMay's doctrine of strategic bombing.

The election of Dwight Eisenhower as president in 1952 brought the first realistic policy review of the nuclear bomber. Eisenhower hired a crusty, straight-spoken automaker, Charles "Engine Charlie" Wilson, CEO of General Motors, to be his defense secretary. One of Wilson's first acts called for a full-fledged look at Pentagon research and development programs, including the atomic airplane. Thus began a long and contentious endgame for the nuke-powered bomber.

Wilson's inquiry revealed that the Air Force and the AEC had not even begun to solve the problem of how to shield flight crews from dangerous radiation. A high-thrust reactor would generate so much radiation that a crew could not operate the plane without risking major radiation doses. The conundrum proved incapable of resolution. A 1961 postmortem in *Science* magazine described the problem well:

> The shield requirement, if power is constant, goes up roughly with the square of the diameter of the reactor. This means that an engine that can be put into an airplane must be driven by a very small reactor releasing a great deal of energy. This meant that, to keep the weight of the shielding down to a point where the plane could fly, reactors had to be built that could operate at temperatures about 500 percent higher than those that would be required in the first atomic submarine, another project begun about the same time. To keep the cost of the plane down to something that would not be entirely unthinkable, these

materials in the reactor had to be able to survive the intense heat and radiation for a reasonably long time.

Radiation shielding was ultimately the physical dagger in the heart of the nuclear airplane program. Air Force, AEC, and GE engineers were never able to solve the problem. Ultimately, the shielding paradox resulted in a steady degradation of the estimates of the capability of a nuclear bomber, which further eroded the political support for the project among military planners and the Pentagon hierarchy.

The Wilson review led the Eisenhower administration to decide to dramatically scale back the flying reactor, focusing on further design while stopping all work on actually building the bomber—a political decision to kill the atomic airplane, albeit in slow motion. Wilson let this nuclear cat out of the Eisenhower administration's policy bag at a June 1953 press conference focused on the Korean War.

Wilson noted that the current Air Force–AEC nuclear aircraft program would have to be redirected because it was sacrificing speed for weight (in other words, the Air Force was backing a slower, less militarily-capable aircraft in order to get it into the air). At a press conference in January 1954, defending the Eisenhower administration's proposed 30 percent cut in the AEC budget, Wilson acknowledged that the nuclear bomber was an illusion. He said, "There's no use letting the Air Force say, we built the first one, unless that first one had definite military utility and was not merely a mechanical and technical curiosity."

"If everything had worked out perfectly," said Wilson, "it still would have been a bum airplane."

But the Eisenhower administration quickly found itself in political distress at the hands of the joint committee, aided by back channels from the Air Force to the committee. While the committee was nominally headed by Republicans in 1953 and 1954, under the chairmanship of Rep. W. Sterling Cole (R-NY), there was strong bipartisan JCAE support for the nuclear airplane and against the Eisenhower position.

The JCAE champion for the nuclear plane was Rep. Melvin Price (D-IL), who regularly put public pressure on the White House and Pentagon. His tool, which he used with practiced skill, was the press.

Born in East St. Louis, IL, in 1905, Price worked for years as a sports writer before getting a staff job with Rep. Edwin Schaefer (D-IL) in 1933. He was elected in his own right in 1944 (and served in Congress until his death in 1988).

Former journalist Price had an easy rapport with reporters in Washington. The JCAE most often met in complete secrecy, but when meetings concluded, Price would hold impromptu press conferences. When the Democrats recaptured the House and Senate in the 1954 election (setting the stage for forty years of Democratic dominance), Price became chairman of the joint committee's research and development subcommittee, where he endlessly promoted the nuclear airplane. He also made sure the Appropriations Committee kept the project funded well above the level set aside in the Eisenhower budgets.

While the Eisenhower administration was inclined to kill the ANP project, the Air Force, pushed the AEC and the joint committee to keep the program alive and named Gen. Donald Keirn, who had been the Air Force liaison to the Manhattan Project, to run the nuclear bomber program, in a joint Air Force–AEC position.

Le May's colorful henchman, Gen. Roscoe Wilson, oversaw Keirn's day-to-day handling of the program and its progress. Wilson and Keirn, while working with the Manhattan Project, had persuaded LeMay in 1946 to launch the A-bomber project at Oak Ridge, granting a basic contract to Monsanto, then the prime Oak Ridge contractor. Appointed to the Manhattan Project in 1943 to oversee the special interests of the Army Air Force, Wilson took Keirn, his classmate at West Point, under his wing. Working as the power plant technology chief at the Wright Patterson air base in Dayton, Ohio, early in the war, Keirn was instrumental in getting Great Britain to turn over jet engine technology to the United States (some accused him of stealing the technology from the Brits). Keirn persuaded Wilson of the need for the nuclear-powered bomber. The Air Force brass initially brushed Wilson off, but he made the sale to Le May in 1946.

While Keirn's dual Air Force–AEC appointment was the same arrangement as the Navy had with Rickover, Keirn was a far different character. The admiral was a genius for publicity as well as possessed of a driving desire to succeed. Rickover was performing brilliantly with the nuclear submarine. The *Nautilus* triumphed in its sea trials in January

1955, while the nuclear bomber crew floundered over their atomic engine technology.[31]

A *New York Times* profile described Keirn in May 1955: "The man behind the atomic plane is almost unknown. He would like to keep it that way." The *Times* added, "At the headquarters of the Atomic Energy Commission, where the general works, officials recently found to their surprise that they lacked any biographical data about him. Yet, he has been connected with the commission since 1946."

Mad-bomber Le May was also keeping up his interest in the atomic plane, now from his position as commander of the Strategic Air Command. Wilson became a key Air Force staff officer shortly after the war, overseeing technology developments. LeMay told the JCAE in 1956 that an early flight of the nuke plane was possible. The next month, Keirn told the committee that there would be a ground test in 1959, with flight in 1960. That was characteristic of the unfounded optimism that pervaded the nuclear bomber program.

Herbert York, a Pentagon nuclear analyst and physicist who began his career at the University of California–Berkeley's Livermore radiation laboratory, working for Edward Teller, wrote that "a review of the technical progress in the program and subsequent budget cuts by the Defense Department led to postponing the flight target date by eighteen months. In December (1956) an experimental reactor operated a turbojet in a laboratory for several hours, but not at a temperature suitable for flight propulsion."

After serving as the first director of the Livermore National Laboratory under the legendary Earnest Orlando Lawrence, York became head of engineering and research at the Defense Department in 1958, an Eisenhower appointee, while only thirty-seven years old. He worked briefly on the White House science staff before his Pentagon appointment. A dozen years later, he wrote, "The political pressure to put a plane in flight as soon as possible eventually proved fatal to the program. The part of the program

---

31 In 1954, Oak Ridge worked on a small, 2.5 MW-thermal molten salt (sodium fluoride) reactor moderated by beryllium oxide for use in the bomber program, later taken over by General Electric, but never developed to the point where it could have powered flight. Oak Ridge's Don Trauger said, "It was called a fireball and it was. It was to run red hot."

which was supposed to develop reactor materials had by no means reached the point where it could be certain of coming up with something suitable." York was a major arms control advocate. He strongly supported killing the nuclear bomber. But it took nearly four years to accomplish the task and didn't happen on his watch.

The Air Force Scientific Advisory Board had repeatedly recommended that the nuclear airplane program should stop pushing for early flight and concentrate on reactor development. In May 1957, another advisory board recommended a low-flying plane, a repudiation of Le May's vision for the manned aircraft. York commented, "The reason for specifying a low-level plane was simple: no one knew how to design a reactor suitable for any other kind of flight."

The political tug-of-war over the nuke bomber continued through the Eisenhower administration, with the president and his top Pentagon advisers trying to slow down the program while the Air Force, the AEC, and the JCAE pushed for rapid progress. In Congress, Mel Price started playing the Russians-are-coming card as often as he could. The Soviet Union had recently put Sputnik in orbit, ahead of the U.S. space program, badly spooking the United States and undermining its self-image as the world leader in all things scientific.

The Russians looked invulnerable. Price repeatedly asserted that the Russians were developing the nuclear bomber, and the United States would be beaten again if it didn't speed up the program. With the Republicans in control of the White House and the Democrats holding Congress, the nuke plane became a political issue, as the Democrats hammered the administration over the Russian atomic bomber program.

No evidence supported the assertion that the Russians had such a program. The definitive history of the Soviet nuclear program, *Red Atom*, by Paul R. Josephson, makes no mention of the bomber program. A February 2006 *Washington Post* obituary of Robert B. Hotz, the legendary editor of *Aviation Week* magazine, observed, "Mr. Hotz had occasional missteps, such as a 1958 story claiming that the Soviet Union had perfected a nuclear-powered bomber—which never existed." Eisenhower, who had access to all the intelligence, said of the *Aviation Week* story shortly after it appeared, "There is absolutely no intelligence to back up a report that Russia is flight-testing an atomic-powered plane."

Nevertheless, the proponents of the nuclear bomber, including Mel Price and a majority on the JCEA, continually voiced the notion that the United States was in a race with the Russians to develop nuclear-powered bombers. As is often the case with Congress, facts never seemed to interfere with the impetus of politics.

There was also concern over the impact of the nuke plane on public health and safety. It would require an enormous runway for takeoff, and the Pentagon decided it could not fly over land, for fear of the consequences of a crash. There was also the issue of what would be blasting out of the rear end of a nuclear-powered jet engine. Convair, one of the major contractors for the bomber, initiated a project to try to characterize the tailpipe emissions from a nuclear jet engine. They called it, apparently with some sense of humor, Project Halitosis.

At the same time, other military aircraft and missile technologies were making great strides, undermining the original mission for the nuclear bomber. Le May's SAC was flying B-47 and B-52 strategic bombers and routinely filling up on jet fuel in flight. The Navy was well on the way to developing solid-fueled sea-based ballistic missiles in nuclear submarines.

Meanwhile, the expectations for the atomic bomber steadily diminished. So-called mission creep, where engineers keep adding functions and capabilities to new weapons systems, is typical of the U.S. military research and development enterprise. The A-plane suffered from mission retreat. In March 1958, *Newsweek* reported, "Last week, Air Force Secretary James H. Douglas said an engineer told him the more one works on a nuclear plane, the more one is convinced 'she will not fly high, she will not fly fast, and she might not fly at all.'"

The Air Force, with no obvious awareness of the irony, was now calling the nuclear airplane project CAMAL, for Continuously Airborne Missile Launcher and Low Level System. While trying to attach itself to the politically potent missile program, the new moniker also acknowledged, in the "low level" description, that the aircraft would be severely limited. The moniker led to a wisecrack in Washington that a camel is a horse designed by committee and a CAMAL is a plane designed by a joint committee.

In the 1960 presidential election, John F. Kennedy, the Democratic nominee, blasted the Eisenhower administration for a "missile gap".[32] With the focus on missiles, the nuclear war bird was rapidly losing political altitude. An article in *Science* in September 1960 described the A-bomber program as "the most controversial in the Defense Department." The article said Price and the JCAE "would like to see a plane in the air, any plane. They are willing to settle for what is called a 'flying platform'—that is, a machine that may have no function beyond demonstrating that it can get off the ground."

Rep. Dan Flood (D-PA), a flamboyant former Shakespearian actor from northeastern Pennsylvania, known for his ostentatious mustache, cape, cane, and temper[33], demonstrated the lowered expectations for a nuclear airplane. At a House appropriations hearing, Flood exploded, "I do not care how big it is, and I do not care how much it costs. I want the Defense Department to propel an airframe with nuclear power fifty feet off the ground, 20 mph if need be. This is a horse race." Of course, it wasn't. It was a CAMAL race, with no real entrants and no prospects that any animal would make it to an airborne finish line.

Kennedy narrowly won the 1960 election against Republican Vice President Richard Nixon. Kennedy's administration immediately focused on their perceived need to expand the missile program to address the alleged gap. York stayed on in the White House for a couple of months to help in the transition. He suggested cutting the nuclear airplane. Kennedy's science advisor, Jerome Wiesner of MIT, also a physicist, quickly agreed.

The White House signaled its intent in March 1961, when the *New York Times* reported, "White House advisers say that President Kennedy has expressed concern and amazement at the cost and time requirements of the project. Since it was started fifteen years ago, slightly more than $1 billion has been spent. The general estimate is that it will take at least $700 million more and up to ten years before a plane can be put into operational use."

---

32 The missile gap turned out to be entirely bogus, but helped Kennedy win the election.

33 Flood was fond of driving around Washington during the temperate months in a top-down convertible accompanied by well-endowed blondes.

The ax fell on March 28, when the Kennedy administration announced its first defense budget, calling for major increases for the Polaris sub-based program and the Minuteman land-based missile. Kennedy proposed killing the B-70 conventional bomber and the nuclear bomber.

Price was apoplectic; he had not only guzzled the atomic Kool-Aid, but had cooked it up from the beginning. Price tried to mount a rear-guard action in Congress, arguing against all evidence that the atomic bird was about to soar. He was unable to challenge the popular, newly-elected president from his own party.

In November 1962, speaking to the Atomic Industrial Forum, the nuclear industry's lobby, Price claimed history would prove him right. "Americans someday will have a nuclear-driven airplane," he predicted. "The program to build one will start again the day after Russia has put a squadron of such atomic-driven ships in the air."

He was wrong about the Russians and wrong about the U.S. bomber to nowhere. Today, some 50 years later, neither the U.S. nor the Russians, the former Soviet Union, have a nuclear powered airplane. Why? Such an airplane makes no sense.

7. The Road to Jackass Flats

A large, remote portion of the Atomic Energy Commission's Nevada Test Site (now euphemistically called the Nevada Nuclear Security Site) is formally known as Site 400, part of the larger Area 25. It is, like almost all of the land in the vast, 1,350-square-mile test site, dry, barren and, most important, remote and unoccupied.

After examining several locales in search of a secure place within the borders of the United States, President Truman in December 1950 designated the Nevada ordnance test area, some sixty-five miles northwest of Las Vegas, for the nation's nuclear bomb tests. The White House publicly announced the selection in January 1951. *Time* magazine reported in May 1951, "With its customary air of guarded caution, the Atomic Energy Commission last week announced that it would begin a new series of tests 'in the near future' at the Las Vegas Bombing and Gunnery Range in Nevada. The site is now on 'permanent' status, said the AEC, and will be used for both atomic and ordinary explosives."

Site 400 is at Latitude 36.854 N and Longitude 116.2926 W, in the south-central portion of the test site. It can be found at the foot of Skull Mountain, next to Death Valley. Among the few locals, and later among the scientists and engineers trying to invent a nuclear rocket engine for manned flights to the moon and Mars and a ramjet for unmanned bomb-tipped missiles destined for Russia, the area was best known by its colloquial moniker: Jackass Flats.

Polymath physicist Freeman Dyson, possessed of the clearest mind and the most lyrical of the scores of prominent scientists intoxicated with the concept of atomic spaceflight, visited Jackass Flats in 1959. He wrote twenty years later: "Jackass Flats was as silent as Antarctica. It is a soul-shattering silence. You hold your breath and hear absolutely nothing. No rustling of leaves in the wind, no rumbling of distant traffic, no chatter of birds or insects or children. You are alone with God in that silence. There in the white flat silence I began for the first time to feel a slight sense of shame for what we were proposing to do. Did we really intend to invade this silence with our trucks and bulldozers, and after a few years leave it a radioactive junkyard?"

The AEC located all three of its major attempts at utilizing nuclear power for missiles and rockets—projects Rover, Pluto, and Orion—at Jackass Flats. Although the ax got wielded in the cool, marbled halls of Washington, DC, the dream of nuclear flight expired in the remote Nevada desert.

Nuclear fuel's great energy density enticed many of the proponents of atomic rocket propulsion. Uranium atoms packed a lot of wallop into a small package, much more than chemical fuels. The difficulty lay in releasing the energy in a controlled fashion. In practice, the atomic beast never got properly unleashed and domesticated enough to push vehicles through the air or space. The nuclear boffins were convinced that chemical rockets were too heavy, too weak, and too expensive to represent an ideal way to explore the skies. But those relatively anemic chemical rockets put men into orbit and on the moon, and powered the workhorse space shuttles for decades.[34]

One postmortem of the atomic rocket program noted, "The advantage of a nuclear rocket is that it can achieve more than twice the specific impulse of the best chemical rockets. For a Mars mission, a 5,000-MW engine would burn less than an hour to provide the necessary velocity for the mission." Nuclear engineer James Mahaffey defines "specific impulse" as "a measurement of the change of momentum per unit of propellant, com-

---

34 There was a stillborn plan to use a nuclear engine in the Space Shuttle, the U.S. space truck that went out of service in 2011.

monly noted in units of seconds." According to Mahaffey, the maximum specific impulse of a chemical rocket engine is 453 seconds. The best specific impulse the AEC rocket scientists were able to achieve at Jackass Flats was 850 seconds in an engine developing 4,500 MW of thermal energy.

The developers of nuclear flight—for air-breathing jets as well as rockets and other space engines—could never clear one major hurdle: the intensely, fatally radioactive exhaust. Still, the prospect of spewing radioactivity over wide swaths of countryside did not deter the nuclear scientists, engineers, and politicians from spending wildly on the experiments.

The AEC and the Air Force approached nuclear rockets with a mind to develop a light-weight, high-temperature reactor to heat a working fluid, generally hydrogen, moving through the core of the reactor. This design, known as a "solid core" rocket, would provide the propulsion needed to push the rocket forward. . A veteran of the Rover rocket program described the basic engine in a 1991 retrospective: "The Rover test reactors utilized a solid core fission reactor. The basic concept employed a graphite-based reactor, loaded with highly-enriched uranium 235. Hydrogen was used as the coolant/propellant due to its low molecular weight. Early tests utilized gaseous hydrogen whereas liquid hydrogen was subsequently used for all tests conducted after 1961."[35]

As in the case for the atomic bomber, the push behind the nuclear rocket engine and its earthly companion, the ramjet missile, began shortly after the end of World War II, amid the red-hot enthusiasm for all things atomic, and as the various government nuclear laboratories and their contractors began looking for work to replace the radioactive gold mine of the Manhattan Project. Nuclear rockets had been an unrealistic dream since before the war. A *Boston Globe* article in May 1940 breathlessly said, "Just add cold water, fly to the stratosphere. One pill of U-235, miracle substance just announced, would drive an automobile for a year." The claims were nonsense, but the enthusiasm was real.

---

35 Scientists and engineers also looked at liquid core designs—which would have mixed the nuclear fuel and the working fluid—and at gas-core engines, with uranium fuel as the gas. Neither liquid nor gas core designs looked practical, even from a theoretical perspective.

In June 1946, the newly-created AEC asked Johns Hopkins University's famed Applied Physics Laboratory outside Washington, DC, for a feasibility study of nuclear space propulsion. Six months later the APL concluded that space rockets were feasible, but the technical obstacles were daunting. The idea languished at the AEC for several years, as the commission faced more pressing matters, such as the need to transition from a single-focus military body to a civilian and military enterprise with multiple goals and myriad projects. Additionally, one early crisis—the Soviet Union's August 1949 detonation of an atomic bomb several years before most inside and outside the government expected—resulted in the crash program to develop the Super, Edward Teller's hydrogen fusion bomb. An official history of the Lawrence Livermore National Laboratory commented, "The Soviet A-bomb changed everything. To Teller (and many others), an American H-bomb seemed the best response to the new Soviet threat."

By the mid-1950s, undeterred by the serious technical challenges the Johns Hopkins report raised, the AEC began to pursue atomic rocketry and nukes in space. A paper by a brilliant young rocket scientist at the Oak Ridge laboratory proved an important milestone in the program. In 1953, Robert W. Bussard wrote a monograph titled "Nuclear Energy for Rocket Propulsion," arguing for the technical advantages of nuclear-powered engines in getting into and exploring space.

With the Bussard analysis giving the agency technical comfort, the AEC began launching its space rocket program. Noted space historian and aerospace engineer Thomas A. Heppenheimer says Bussard's paper "stirred interest, and led to the initiation of an experimental effort called Project Rover at Los Alamos, New Mexico."

The AEC launched Project Rover in 1955 under the direction of Los Alamos. Following the successful testing of the hydrogen bomb in 1951, the New Mexico lab began broadening its base of scientific and engineering activities beyond explosives into what one contemporary described as "other areas of national interest," including rocket engines. As was customary for AEC programs, Los Alamos contracted with major firms to carry out the work on Rover. The early contracts were with Aerojet General, for its rocket expertise, and Westinghouse, a source of much experience on nuclear issues.[36]

The AEC also hooked up with the newly-named Air Force (formerly the Army Air Corps) on the Rover project, with the goal of tactically useful missile. The Air Force was a willing partner, but was placing most of its atomic bets on the A-plane, consistently doubling down on the bomber to nowhere. Ultimately, the Air Force lost both bets.

The U.S. program had its roots in German rocket technology. Germany pioneered the idea of missiles as military weapons with the primitive Fieseler Fi-103, or the V-1 Flying Bomb, also known as the "Buzz Bomb" or the "Doodlebug.—essentially an unmanned and largely uncontrolled air-breathing pulse-jet plane flying at low altitude and designed to crash into random targets.

The military virtue of the V-1 was that it was cheap to make, equivalent to what it cost to produce a Volkswagen car in Germany. Thus, the Germans could make a lot of the missiles and throw them at the only reasonably close target, Britain. Over the course of five months, from June to October 1944, Germany had flown nearly ten thousand Doodlebugs willy-nilly at British targets, mostly in southeast England.[37] The V-1 was a crude weapon designed for inspiring fear, rather than a weapon serving military tactical aims. It flew low and slow, closer to what we think of in modern rocketry terms as cruise missile.

The V-1 was not a true ballistic missile—defined as a missile that, after launch, follows the laws of gravity and motion that apply to falling objects such as shells or bullets fired from guns.[38] When the ballistic missile has burned off its fuel and reached the peak of its trajectory, it is largely on its own—with onboard guidance, but without external directions on how to hit its target.

---

36 Pittsburgh-based Westinghouse had successfully designed and built the power plant for the Nautilus nuclear submarine, the pride-and-joy of both the AEC and the Navy.

37 When the Brits overran the German site firing the Buzz Bombs, the Germans retargeted, firing another twenty-five hundred at Antwerp and other Belgian targets. In total, the German V-1 killed or wounded nearly twenty-three thousand people, mostly civilians.

38 Hence the term "ballistic."

Once the missile is in its bullet mode, it follows a path, or azimuth, drawn by the laws of physics. Webster's defines "azimuth" as "an arc of the horizon measured between a fixed point (as true north) and the vertical circle passing through the center of an object, usually in astronomy and navigation clockwise from the north point through 360 degrees."

The ominously-named *Vergeltungswaffe-2* (Reprisal Weapon–2), commonly called the V-2, succeeded the Buzz Bomb. This weapon, designed by a team of true rocket scientists led by the legendary Wernher von Braun and located at Peenemünde in the Balkans, was based on early rocket science by the American Robert Goddard and was a true ballistic missile.[39]

The V-2 was a spectacular scientific and engineering accomplishment, tainted by the Nazi regime that invented it and produced it using slave labor. The missile lifted a one-ton explosive warhead, powerful enough to destroy a city block, 60 miles high and 250 miles downrange at 3,300 mph (nearly four-and-a-half times the speed of sound) to its target. It even had a primitive guidance system using a gyroscope to add precision to its ballistic course.

The V-2 was designed to be a game-changing technology for Germany, which, by the fall of 1944, was clearly on the verge of losing the five-year-old war. Perhaps, if the missile had been developed earlier in the war, it could have transformed the face of the war. But the V-2 came too late and ate too much of Germany's stressed and dwindling resources.[40] Despite their tremendously advanced technology, the V-2s were not very effective. While their speed and altitude of attack made them invulnerable to anti-aircraft weapons (unlike the low and slow V-1), the V-2s were notoriously inaccurate. The V-2 was also expensive, costing as much as a four-engine conventional bomber, which had a greater payload of destruction and was slightly more accurate.

Freeman Dyson wrote a typically charming assessment of Germany's V-2 campaign in his 1979 book, *Disturbing the Universe*. Dyson wrote that the V-2 was a "technically brilliant" device that

---

39 The namesake for NASA's Goddard Space Flight Center in Maryland.

40 Through March 1945 (the war ended in May), the Germans fired off some thirty-one hundred V-2s, most of them at targets in the Netherlands (over sixteen hundred) and England (about fourteen hundred).

made no economic or military sense. I became aware of the success of the Peenemünde project in the fall of 1944, after the V-1 bombardment of London had ended, when I heard the occasional bang of a V-2 warhead exploding. At night, when the city was quiet, you could hear after the bang the whining sound of the rocket's supersonic descent. At that time in London, those of us who were seriously engaged in the war were very grateful to Wernher von Braun. We knew that each V-2 cost as much to produce as a high-performance fighter airplane. We knew that the German forces on the fighting front were in desperate need of airplanes, and that the V-2 rockets were doing us no military damage. From our point of view, the effect of the V-2 program was almost as good as if Hitler had adopted a policy of unilateral disarmament.

While the V-2 failed to rescue the doomed Nazi regime, the Allied victors well understood the future importance of German rocket wizardry. The fall of Nazi Germany set off a scramble between the Soviets and the Allies to seize the remnants of the German program. The United States mounted Operation Paperclip to capture as many V-2 missiles and parts as possible, as well as the key scientists. Von Braun and many of his colleagues decided they would rather be captured by the Allies than the Russians or assassinated by their Nazi overseers to prevent their capture. The von Braun team went into hiding until the fighting was over. He then arranged to turn himself and much of his team over to the United States. He subsequently became a leading light in the U.S. missile program. The United States also seized some three hundred trainloads of V-2 missile and parts. With seizure of the German hardware and, more important, the wetware represented by von Braun and others, the U.S. missile program catapulted into existence.

In the early days of U.S. military rocketry, a fierce rivalry broke out between the Army and the Navy over missile development. Both services (the Air Force was still part of the Army) vied for supremacy in developing land-based, ballistic missile strike capability, based on the liquid fueled V-2 technology. The Army program was built around von Braun and headquartered at the Redstone Arsenal in Huntsville, Alabama.

When the Air Force split off from the Army and became a separate service in 1947, the former siblings and now rivals began squabbling over rockets and missiles. The Air Force argued that missiles should be part of its strategic mission, built around long-range bombers such as the B-29, which dropped the atomic bombs on Hiroshima and Nagasaki, and the later six-engine, propeller-driven B-36 behemoth. The Air Force constructed its program around the brilliant mathematician John von Neumann and the Atlas missile, a V-2 spinoff. The Consolidated Vultee Aircraft Corp., also known as Convair, was the chief Air Force contractor, and the service had a missile launch site at Cape Canaveral, Florida.

The Navy was interested in submarine-based missiles, something the Germans were exploring when the war ended. But the Navy was also putting money into land-based missiles, centering its program at the Naval Research Laboratory outside Washington, DC, around the ill-fated Vanguard rocket. The liquid-fueled Vanguard had the imprimatur of the Eisenhower administration as the foundation of the nation's missile program. Early in the 1950s, the Navy also forged a strategic alliance with the Army to share the Army's Jupiter missile technology as the basis for the submarine-launched ballistic missiles.

The development of Teller's H-bomb gave great push to missile development. The thermonuclear weapon promised a smaller, lighter warhead that could deliver far greater destruction in an easily handled and lifted package. An intercontinental missile tipped with the threat of national destruction promised to be the ultimate weapon of war.

The Air Force, betting heavily on the atom to power its next generation of bomber, decided to make a parallel wager on unmanned atomic flight, with atom-powered rockets to hoist ballistic missiles and ramjets for low-altitude missiles. Thus, the alliance arose with the Atomic Energy Commission in the Rover and Pluto programs. But by most accounts, it was a second-class Air Force effort, located organizationally in the Air Force's Aircraft Nuclear Propulsion Office, the same institution running the high-profile atomic bomber project.

On October 4, 1957, the world changed—the Soviet Union announced to a surprised world (although it was no surprise to the Eisenhower administration) that it had successfully put into Earth orbit a two-hundred-pound metal beach ball called Sputnik. Its beeping elliptical path around the planet, taking ninety-eight minutes per circuit, shocked and fascinated the world and set off a furious space race between the United States and the

Russians. Consequently, it propelled the AEC's atomic rocket program to national priority.

Before Sputnik, the National Advisory Committee for Aeronautics (NACA, chartered in 1915) managed the U.S. non-military space flight effort. Both the Russians and the United States were attempting to launch an Earth-orbiting satellite during the 1957–1958 International Geophysical Year. The U.S. effort, run by NACA, was built on the Navy's three-stage Vanguard. That rocket demonstrated the feeble nature of the U.S. missile program by spectacularly exploding in December 1957, having been rushed in an attempt to emulate the Soviets.

The Eisenhower administration and Congress turned the largely low-key NACA into the National Aeronautics and Space Administration on October 1, 1958, almost exactly a year after Sputnik. NASA inherited much of the Air Force and Army missile programs, and saw the atomic rocket engine as well-suited for long-haul space missions to the moon, Mars, and beyond. NASA soon began discussions with the AEC over the atomic engine.

In essence, the Sputnik launch turned the military rocket program aimed at the Russians into a large civilian-oriented space race, although the military component remained important. In addition to the well-funded Rover program, the Atomic Energy Commission was also working on the Orion project, a plan to use explosions to push manned vessels into and through space, and the diabolical Pluto project, a planned hypersonic, low-flying, H-bomb-tipped cruise missile designed to irradiate and destroy everything in its path. Rover, Orion, and Pluto never made it out of Jackass Flats.

8. Flightless Birds and Flying Elephants

While Sputnik didn't surprise the intelligence agencies in the Eisenhower White House, the public viewed the Russian success as a major American technological defeat. Atomic enthusiasts saw an opening presented by the creation of the new, high-visibility space agency. The Rover program had been sitting on the Air Force–AEC back burner for two years. Under the direction of the Los Alamos National Laboratory, the nation's chief weapons development center, Rover program scientists and engineers had been working on early research and development of a nuclear rocket engine. Their goal was to develop an engine that would produce temperatures in excess of 2,200 K.[41]

Eisenhower's presidential executive order creating NASA[42] also transferred all of the Air Force's interests in the space rocket program, along with some $58 million in funds, to the new civilian space agency. The order specified that the transfer included "projects of the Advanced Research Projects Agency and of the Department of the Air Force which relate to space activities (including lunar probes, scientific satellites, and super-thrust boosters)." This was the basis of what became, under NASA, the Nuclear Engine for Rocket Vehicle Application (NERVA)[43] program.

The AEC and NASA turned to Project Rover for a space propulsion program they hoped would mimic the Navy–AEC collaboration on the successful nuclear submarine. The Air Force and the AEC had already used

---

41 Kelvin is a temperature scale used in physics that starts at absolute zero, the point where all thermal motion stops, using the same temperature intervals as the Celsius temperature scale, so 2,200 K is about 1,900 degrees Celsius.

42 E.O. 10783

43 Could our government function without acronyms?

the Rickover model for its ill-fated atomic-powered bomber project, which was about to die soon after the rocket program was to begin. After conceptual and engineering designs were completed, Rover/NERVA came to life at Jackass Flats in 1959. At that point the A-plane was clinging to bureaucratic life by its political fingernails, and the grip would soon slip; in eighteen months the new president would both cancel the bomber and announce a national goal of putting a man on the moon by 1970.

The Rover program ultimately fared no better than the A-plane, and its history eerily mirrored the terrestrial project. Project Rover aimed to use the power of the atom to put men into space. Its key was an engine humorously named Kiwi, for the flightless, nocturnal bird of New Zealand. The nuclear kiwi, indeed, was not designed for flight and never got off the ground. After fifteen years of effort and an expenditure of $1.5 billion dollars,[44] Washington pulled the plug on the atomic-powered rocket. International arms control politics and NASA's success with chemical propulsion doomed the atomic venture.

The program combined the AEC's reactor research and development with NASA's practical engineering experience in space flight, and aimed to produce a vehicle that could move humans from a terrestrial launch to the moon. The AEC focused on the rocket engine. The job of putting the nuclear engine into an actual missile that could fly landed at NASA's Marshall Space Flight Center, in a project known as Reactor in Flight Test (RIFT). The idea was to mate a manned space probe powered by the atomic engine to a chemically-fueled Saturn V rocket. The Saturn would lift the nuclear-powered vehicle into orbit, avoiding launch pad radiation, a clear problem for nuclear-powered rockets. From there, the atomic rocket would push the manned space capsule to Mars. The 1961 plan called for a March 1981 launch and an August 1982 Mars touchdown.

At the time, Marshall officials told Congress, "Possible applications of a nuclear Saturn would include carrying large payloads into Earth orbit, to the moon, or beyond. With chemical rockets, only about 5 percent of the total weight is payload. With a nuclear system, 16 percent can be payload."

---

44 Dollars not adjusted for inflation.

In the charming 1956 book *The Exploration of Mars*, authored with noted German expat rocket science and science writer Willy Ley,[45] von Braun said, "It is entirely possible…that within a decade or so successful tests with some sort of nuclear rocket propulsion system might be accomplished." The rocket scientist added, "But the chances are that nuclear rocket propulsion systems will find their first application not in ground-launched rockets but in deep-space rocket ships."

The formal alliance between the Atomic Energy Commission and NASA began in 1960. In 1961 the AEC created the Space Nuclear Propulsion Office, headed by a bright, personable thirty-seven-year-old NASA physicist, Harold B. Finger.[46] Finger started working for NACA in 1944 at the Lewis laboratory in Cleveland. After training in nuclear engineering, he moved to Washington in 1958 to oversee NASA's interests in nuclear engines. Finger became the first, and only, head of the joint AEC–NASA space propulsion office.

The Kiwi predated the marriage of NASA and the AEC, with the first flights at Jackass Flats taking place in July 1959. The primitive machine, firing a superheated, super-radioactive exhaust stream out of its rear end, was an ungainly conglomeration of tubes, pipes, flaps, and bits of metal, standing on the end of spidery scaffolding.

The obstacles that these and all nuclear rockets had to overcome involved the impact of the forces of atomic energy on the materials used to house those primal powers. Heat was required, but also an enemy. To provide enough hot gases to power a rocket, the nuclear reaction had to be close to the melting point of the fuel and the metals and materials containing them. Vibration was another problem. The enormous power of splitting atoms and turning the energy into propulsion produced prodigious shakes, rattles, and rolls. Those seeking to use those forces had to work hard to keep their machines from flying apart at the seams.

The Kiwi engines featured graphite as the main structural material, as well as an important element in sustaining the nuclear reaction. Graphite is

---

45 Ley was von Braun's childhood friend and fellow youthful rocketeer, who fled Germany and the Nazis in 1935.

46 In the 1980s and early 1990s, Finger headed up the U.S. nuclear power industry's Washington lobbying group.

a pure carbon solid, one of two "allotropes," or crystal forms, of the element. Graphite has long been used in nuclear projects for a couple of important reasons. First, it is a neutron "moderator," which means it can slow down neutrons as they fly out of a split atom of uranium, making it more likely that the neutron will slam into another uranium atom, producing more neutrons. This increases the chain reaction, requiring less uranium to sustain the fission of the uranium. Graphite is also very strong at high temperatures and actually gets stronger as temperatures increase.

But there are downsides to graphite. First, when it comes into contact with hot hydrogen gas—which was the coolant and propellant in the nuclear rockets—graphite erodes quickly. The other problem is that it can catch on fire, although not easily. As many have observed, graphite is simply a very high grade of coal, a step above anthracite, or hard coal. The British discovered the problems of graphite fires the hard way, at their Windscale plutonium production pile on October 10, 1957. The graphite moderated reactor inexplicably caught fire. By the time firefighters and nuclear engineers got the blaze under control, the severely damaged plant had produced two large releases of radioactivity into the British countryside.

Engineers and scientists are wont to put the best face on their experiments. One of the truisms of the scientific endeavor is that even failures teach valuable lessons. That was true for the nation's atomic energy research, including the Kiwi tests. The first Kiwi test, Kiwi-A, used plates of highly-enriched uranium dioxide as fuel. The reactor reached a temperature of about 2,700 K at a thermal power level of 70 MW. The scientists defined it as a success, for having "demonstrated the principle of nuclear rockets." Unfortunately, noted a postmortem, "Vibrations during operations produced significant structural damage in the reactor core."

Kiwi-Á followed in a matter of hours, using a different design for the fuel elements. Unlike the uncoated plates that contained the fuel in the first test, the second reactor had UO_2 cylinders encased in graphite, held in channels coated with niobium carbide, a high-temperature ceramic. The machine ran for six minutes, reaching on output of 85 MW thermal (MWt).[47] An engineering assessment commented drily that "some

---

47 A thermal megawatt is not the same as the way we normally see the term "megawatt" used. In conventional usage, megawatt refers to the measure of electrical power produced

structural damage occurred in the improved design during its six-minute test." The next Kiwi test came in October 1959, with the Kiwi-A3 reactor, again using a slightly different fuel array. The conclusion: "Some core damage occurred during the five-minute test…which reached power levels of 100 MWt, with some fuel elements showing blistering and corrosion. Generally this reactor test was considered successful." Note the use of the passive voice. Who considered it successful? The engineers, no doubt, but it looks a lot like a Pyrrhic victory.

The AEC's Raemer Schreiber, who oversaw Jackass Flats projects for Los Alamos, told a *Time* magazine reporter about the first Kiwi test, "The engine worked perfectly." *Time* was among the great majority of U.S. publications at the time who acted as tame scribes for the atomic-industrial complex. In the 1959 report on the Kiwi-A tests, *Time* gushed that "Kiwi strutted its stuff." Maybe it could strut, but it certainly couldn't fly.

The list of Kiwi "successes" is astonishing. Kiwi-B1A's test came in December 1961, with the aim of reaching 1100 MWt. At 300 MWt, thirty seconds in, the engineers cancelled the test "due to a fire caused by a hydrogen leak in the reactor exhaust nozzle." In September 1962, the engineers reran the December 1961 test, with the Kiwi-B1B machine. The run "was terminated within a few second when several fuel elements were ejected from the reactor exhaust nozzle." In short, the reactor blew its guts out through its atomic anus. Of these early Kiwi tests, AEC–NASA nuclear rocket czar, Harry Finger, commented off-handedly nearly forty years later, "Problems showed up."

In a 2000 retrospective at the American Nuclear Society international meeting, Finger focused on the experience with the Kiwi-B4A test in November 1962. Again, the machine was configured with new fuel elements in a new geometry. As with all prior tests, the Jackass Flats rocket crew had hopes the test would provide the basis for a later engine that could fly. "However," Finger said, "the high expectations for that November 30, 1962 test were quickly turned off when the test started and flashes of light in the nozzle exhaust indicated core damage as the power increased over 250 MW."

---

in a power plant. Because of conversion losses, 2,000 MW thermal equals about 650 electrical MW.

The timing of the failed Kiwi-B4A test could not have been worse. The bomb site boffins probably thought they had been smart in scheduling a reactor burn the week before President Kennedy, Vice President Johnson, and a coterie of followers showed up at the test site. The DC contingent appearing at the Jackass Flats show on December 8 included: AEC Chairman Glenn Seaborg, White House science advisor Jerome Weisner, NASA chief Robert Seamans, White House national security advisor George McBundy, Defense Department research chief and Lawrence Livermore National Lab veteran Harold Brown, Kennedy press secretary Pierre Salinger, and a host of lesser lights. The Washington entourage had earlier spent a day at Los Alamos, getting briefed on the broad range of the lab's programs.

While the dignitaries were getting the typical snow job at the Test Site, said Finger, the failed Kiwi-B4A reactor had been moved to the maintenance and disassembly building, known to the acronym-addicted rocket crew as the "MAD building." But the technicians hadn't disassembled the engine "so we did not yet know what problem had actually occurred." Finger had told the distinguished Washington guests that something untoward had taken place at the engine stand but didn't know the extent of the problems. "On disassembly," Finger said, "it was found that almost all of the fuel elements had been broken as a result of severe vibrations that had been experienced through the entire core."

What followed from the Kiwi-B4A failure was a classic bureaucratic knock-down-drag-out donnybrook involving Finger and Los Alamos lab director Norris Bradbury, each fully armed with technical staff, along with cheering sections from NASA's Lewis center in Cleveland, Ohio, and Alabama's Marshall space flight center. Finger laid down the law: no hot reactor testing "until we had gone through thorough work to identify the causes of the failures and to develop well-defined solutions to those problems." Finger recalled that Bradbury, reflecting the full-steam-ahead psychology of the Los Alamos bomb builders, "objected, saying that I would kill the program if there was no continued reactor testing. I responded that contrary to his position, there was no question we would kill the program if reactors continued to have major failures." Finger won, and Kiwi retreated to the engineering drawing board.

A final Kiwi test, in January 1965, proved the most "successful" of all. The droll engineers at Jackass Flats called it Kiwi-TNT. The technicians tethered the Kiwi-B4 reactor to the test stand and let it run wild, without any control of the nuclear reactor, known in the trade as a fast excursion. According to the scientists, the purpose of the test was "to confirm theoretical models of transient behavior"—in other words, let's see if we can blow it up. The reactor—for the first time—performed as advertised and blew up in a spectacle of flame, spewing debris hundreds of feet in the air: mission accomplished.

In its 1959 article on the "successful" first tests of Kiwi, *Time* magazine commented,

> After a few days, when radioactivity dies down somewhat, the unshielded reactor will be hauled along a railroad track by a remote-controlled locomotive to a special MAD (Maintenance, Assembly, and Disassembly) shop, where mechanical hands will take it apart. The condition of its still deadly interior parts (examined by periscope, TV, or through thick, transparent shields) will tell the Los Alamos men much about how to build nuclear rockets that actually fly. The code names for them are ready: Dumbo and then Condor.

As the boys at Jackass Flats were getting ready to test the first Kiwi power plant, a parallel design project was underway in New Mexico, at Los Alamos, code-named Dumbo. Named for Disney's adorable flying elephant, it was abandoned in 1959. About the same time, the Rocketdyne Corp. of Los Angeles did an analysis for Los Alamos of the costs of atomic-powered spaceships versus chemical propulsion, using a hypothetical, generic atomic engine Rocketdyne called Condor, the nearly-extinct California carrion-eating bird. Neither Dumbo nor Condor ever came to life.[48]

The AEC nuclear propulsion program resumed hot testing rocket engines in the summer of 1965, with completely new designs. The new engines, ambitiously dubbed the Phoebus (Greek for "sun") series, performed far better than the first attempts. The most promising test, Phoebus

---

48 The folks at the AEC weapons lab were fools for funny names for projects. One Los Alamos reactor was named Topsy, and one engineer commented, "She just growed." Another Los Alamos project was Lady Godiva, although the relationship between a nuclear reaction and a naked woman on a horse was never clear.

2A, came at Jackass Flats in June 1968. This turned out to be the most powerful nuclear reactor ever tested, producing 5,000 MWt. It operated for twelve-and-a-half minutes at a temperature of up to 2,310 K, just short of the melting point of the fuel.

Phoebus's fine performance came too late to rescue the already-stumbling nuclear rocket program. The United States was having great success with chemical rockets—headed to the moon by the end of the decade, as President Kennedy had promised in 1963. At the same time, international agreements limiting nuclear weapons testing and nuclear fallout were on the horizon, causing great anxiety and consternation across the wide-spread atomic weapons endeavor. Finally, and perhaps most significantly, the war in Vietnam was beginning to stress the U.S. budget, and White House bean counters started looking seriously at plucking radioactive legumes.

In his 2000 retrospective, Finger ruefully commented, "Unfortunately, as many commentators at the time and historians of the space program have written, the technology of nuclear rocket propulsion was fully demonstrated as ready for flight mission applications, but the deep space missions whose accomplishment depended on nuclear propulsion applications were not part of the U.S. space program nor were any such missions planned. In addition, budget pressures led to the shutdown of the program in 1972."

9. The Devil Flies Nukes

The AEC–NASA Rover rocket program had a nasty neighbor at Jackass Flats, just over the horizon from the tethered, flightless Kiwi. This was the aptly-named Project Pluto. Unlike the relatively benign Rover program, Pluto was a truly diabolical joint AEC–Air Force military effort. Its aim was to spread death and destruction as widely as possible over enemy territory, a worthy expression of the strategic vision of Curtis LeMay. One historian of nuclear power called it "the nastiest weapon ever conceived." Fortunately, conception did not result in birth.

Pluto was named neither for the ninth planet in the solar system nor for Mickey Mouse's friendly pooch, but for the Greek god of the underworld (and also namesake of the radioactive, fissile element plutonium). Georgia Tech's James Mahaffey called it "the weapons system from hell."

A direct descendant of the low-and-slow German V-1, this upgrade was to be a hypersonic (3,000 mph), low-flying, radiation-spewing, nuclear warhead-toting, pilotless cruise missile with a mission to destroy Moscow and other sites in the Soviet Union. Pluto was the most secret of an array of secret U.S. atomic weapons programs. Pluto was conceived as another arm of a massive nuclear first-strike capability aimed at keeping the Soviet Union hemmed in by the U.S. Air Force, a companion to the atomic-powered bomber. It was also part of a plan to continue the postwar ascendancy of the Air Force among the three military services.

Gen. Donald Keirn, the Air Force lifer who ran the nation's military nuclear flight program with a dual appointment in the Pentagon and at the Atomic Energy Commission, described the service's glowing goal in a 1960 paper: "To grasp the portent of nuclear power within the Air Force, visualize a fleet of nuclear-powered bombers continuously airborne around the periphery of a would-be aggressor or a force of supersonic, low-altitude ramjet missiles on ceaseless mobile ground alert within the borders of the United States. In this day of advanced technology these concepts are just being fully understood. Yet as far back as 1944 there were men in the Air Force who foresaw the possibilities."

Pluto was officially a Supersonic Low-Altitude Missile, also known within the weapons hierarchy as SLAM. The first description of the concept came in a still-secret paper out of the Los Alamos weapons lab, the birthplace of flying nukes, in October, 1948. The title was "Self-flying Atomic Bombs or the New Mexico Jumping Bean."[49] Despite the secrecy, atomic entrepreneur Frederic de Hoffmann (founder of General Atomics, the creator of the Orion nuclear space project) let the radioactive lion cub out of the bag in January 1949 in a Los Alamos document[50] titled "Minutes of an Informal Meeting on Nuclear Rockets."

Pluto was not a rocket-powered weapon. The heart of Pluto was a nuclear-powered ramjet. This is an utterly simple machine that produces an enormous amount of energy and thrust, but requires air, making it unsuitable for space travel. Rockets provide their own propellant to push out the back of the engine for thrust, meaning they don't need an atmosphere to function. In a ramjet, the engine is pushed forward so air is drawn in, heated to make the air expand, then exits the jet engine in a tremendous burst. Calling it "admirably innocent of moving parts," Mahaffey describes the ramjet as "simply a big air nozzle with a nuclear reactor in the middle of it. Air was crammed into the front of the engine by the leading shock wave, was heated in the white-hot reactor core, and left at higher energy out the back."

Ramjets had been around since 1913, when a French engineer published a paper describing a "flying stovepipe." German scientists experimented with ramjet propulsion in 1939, but switched to pulse-jet technology for the V-1 unguided missiles. The two types of jet engines are similar, but the ramjet is simpler, with no moving parts. The ramjet's shortcoming is that it has no ability to start from rest. It needs a big push to get air moving through the engine. The pulse-jet engine is able to provide its own startup and takeoff.

---

49 Report LA-714

50 LAMS-836

82

As a way to deliver weapons, Pluto also was admirably simple. It was designed to combine high speed, over three times the speed of sound (also known as Mach 3), with low altitude, so it could avoid enemy radar and swoop in so fast that enemy anti-missile measures would be handcuffed. In Stanley Kubrick's seminal 1964 movie *Dr. Strangelove*, actor Slim Pickens, as B52-Stratofortress bomber pilot T.J. "King" Kong, describes how the plane will attack Russia by coming in low and fast. "They might harpoon us," says Kong, "but they dang sure ain't going to spot us on no radar screen." The simplicity of the concept meant that Pluto would have to be a robust machine that could withstand a number of fierce physical forces both inside and outside the airframe. It had to be what one project engineer called "about as durable as a bucket of rocks." Physicist Theodore Merkle, the father of Pluto's infernal reactor, called it "the flying crowbar."

Pluto would have been a versatile weapon of mass destruction. It would have carried up to two dozen hydrogen-bomb warheads to its targets, guided by a ground-following guidance system. The missiles would be launched upward over the target, falling to the ground in ballistic fashion. Beyond the warheads, simply slamming into enemy territory, as its German Buzz Bomb predecessor did, would have produced a large impact and spread radiation widely, on the scale of the 1986 Chernobyl nuclear power plant explosion.

On top of that, Pluto's flight would have spread terror far and wide. It was hardly a stealth weapon. First, it would have been an enormous bird of prey: eighty-eight feet long, weighing sixty-one thousand pounds when launched by a series of conventional chemical rockets to get the jet up to the speed where air was being rammed in. It would have been the equivalent of a flying locomotive.

Pluto would have been incredibly loud coming in at tree-top level, extremely hot, and intensely radioactive. Aerospace engineer John Terry White noted, "SLAM's shock wave overpressure alone (162 dB) would devastate structures and people along its flight path. And, if that were not enough, the type's nuclear-fueled ramjet would continuously spew radiation-contaminated exhaust all over the countryside." Indeed, it was designed to be a devilishly diverse death machine, indiscriminate in the rain of radiation and force of explosion it brought down on its target and in its path.

The Air Force and the AEC decided in 1955 to move ahead with Project Pluto, choosing to keep it separate from the parallel Rover rocket program and the atomic bomber. Congress and the executive branch kept Pluto independently funded, with much of the details of its support out of sight in the so-called black budget. As the Eisenhower administration prepared to unveil the project, Democrats on the Joint Committee on Atomic Energy saw a partisan opportunity. In January 1958, they launched a series of floor statements in the House and Senate calling for firm legislative branch sovereignty over nukes in space. The *New York Times* reported, "Getting a jump on other Congressional committees, the Joint Committee moved to stake out a claim for Congressional jurisdiction over outer space." The Democrats accused the Republican White House of trying to skimp on projects Rover and Pluto.

While Los Alamos ran the rocket show, the Air Force and AEC chose Los Alamos's bitter rival, Livermore, for the ramjet. Livermore picked the brilliant and iconoclastic Merkle to run its adventure in nuclear flight. Merkle earned his doctorate in physics at nearby University of California–Berkeley, Livermore's parent, and went to work at E.O. Lawrence's radiation lab in 1953, shortly after Edward Teller succeeded in persuading the AEC to turn Livermore into the nation's second nuclear weapons lab. When Livermore created its nuclear propulsion group in 1955, it named Merkle to run the show.

Merkle was no plodding atomic bureaucrat. He was a thrower of furniture and breaker of crockery, as described by historian Gregg Herken. One contemporary called him a "bull in a china shop." A co-worker more creatively characterized Merkle's management style: "He wasn't interested in untying the Gordian knot. He cut it." Merkle would not find much favor in today's world of soporific PowerPoint presentations, which graphic guru Edward Tufte argues dumb-down "the analytical quality of serious presentations of evidence." Merkle eschewed "canned briefings" that scientists routinely trotted out to give to policymakers. Livermore colleague Richard Werner recalled, "You use chalk and you talk off the top of your head because you know it. He had no patience whatsoever with people who didn't know how to do things."

The scientists and engineers in Merkle's atomic skunk works faced daunting technical obstacles, heat chief among them. The reactor had to be as hot as possible to get the most energy out of the engine, pushing the airframe forward. But the engine also had to be as small and light as possible so the aircraft could actually get off the ground. That meant it had to run very hot indeed. "The hotter, the better" is the mantra of the ramjet. The engine had to run for hours or days at 2,500° F. At 2,650° F, the aircraft would catch on fire.

At those temperatures, keeping the nuclear fuel from turning into a useless gas was also a major problem. But a Golden, Colorado-based firm that specialized in refractory ceramics came to the rescue. Coors Porcelain Co. agreed to make half-a-million pencil-shaped ceramic elements to hold the fuel and prevent it from vaporizing.[51] As Merkle and his crew worked on the Pluto reactor—known by the code name Tory— they concluded that they needed a large, remote spot to test an Earthbound version at full throttle. Naturally enough, since Livermore was intimately acquainted with the Nevada Test Site, the locus for the static tests turned out to be Jackass Flats, with an eight square mile section devoted to hell on Earth.

Working with an engine as hot (thermally and radioactively) as Tory called for special accommodations. An automated railroad would have to take the hot reactor from its tethered firing site to the MAD building, also known as the hot shop, where technicians could take it apart with remotely controlled, robotic tools and inspect it by three-dimensional television and through thick, radiation-proof ports. The Livermore boffins would watch Tory pulse and strain against its bonds as it spewed enormous amounts of heat and radiation from a shed far away from the test stand and featuring a fully-equipped fallout shelter.

Livermore hired Marquardt Corp. to help plan and design the Jackass Flats facility. Marquardt, founded by aircraft pioneer and barnstormer Roy Marquardt, whom one biographer called "a mix of Tom Sawyer, Tom Swift, and Harry Potter," was a specialist in ramjet technology. The

---

51 That company, now separately-owned CoorsTek, was spun off from the famous beer company, which started a glass foundry in the late nineteenth century to make beer bottles, and eventually moved into ceramics. But it would be a flight of fancy to call Pluto a beer-powered ramjet.

California-based company, formed in 1944, was the major developer of the ramjet engine for the BOMARC missile, the Air Force's surface-to-air missile of the 1950s. Marquardt became a prime Pluto contractor.

Pluto's MAD building had eight-foot thick walls. Erecting the building required the AEC to buy a gravel mine to supply the aggregate for the cast concrete to contain the radiation.

A ramjet needs copious amounts of air rammed into the front of the engine. In Pluto, that would have come by launching the unguided missile with chemical jets until it was gobbling enough air for the nuclear reactor to sustain flight. That can't happen on the ground. So Livermore had to come up with a method of providing hot high-pressure air to the immobile Tory, tethered to the ground for testing. Marquardt calculated that an engine test of twenty-five minutes at full power would require 4.2 million pounds of air at 3,600 pounds per square inch of pressure.[52] To store the pressurized air required an underground steel cylinder thirty feet across and three hundred seventy feet tall. Livermore borrowed pumps from the Navy's Groton, Connecticut, submarine base to stuff the air into the tank. It took twenty-five labyrinthine miles of steel oil well casing to deliver the air to the mouth of the air-hungry jet. The air handling system accounted for $8 million of the nearly $20 million put toward construction of the Pluto test bed.

Pluto escaped the engineering drawing board and entered the corporeal world on May 14, 1961, with the firing of the Tory-IIA engine at Jackass Flats. The ramjet, shining bright red with high-temperature automotive manifold paint, ran for only a few seconds, but established that the superstructure around the machine to provide it with high-pressure air worked: the engine would ignite. Project officials naturally claimed a full success, but it's hard to credit that, given the infinitesimal duration of the test, and what followed: Tory-IIB, which never even made it onto the railroad flatcar at Jackass Flats. The engineers were simply baffled by the complexity of the forces they were trying to tame and had to go back to the drawing board.

So the designers went back to basics, taking three more years to come up with Tory-IIC. In May 1964, the new engine roared for five minutes at 513 MWt. The test was so loud, said engineer James Mahaffey, "You could

---

52 Normal atmospheric pressure is 14 pounds per square inch.

practically hear it running in Idaho." Again, the Livermore engineers proclaimed the test a success and toasted it prodigiously at a local bar. Merkle and his crew were soon talking about a design for Tory-III.

Unbeknownst to the boffins in Nevada, Pluto was running out of support in Washington, and, more importantly, across the Potomac in Virginia at the Pentagon. Since the early days of the project, Defense Department officials, starting with Herbert York in the Eisenhower administration and continuing with Harold Brown of Kennedy's Pentagon team, had questioned the worth of the Pluto project. In a summer 1961 hearing before the Joint Committee on Atomic Energy's research subcommittee, Brown defended proposed funding cuts on the grounds that the project had no military mission that distinguished it from far simpler, cheaper, and more developed technologies.

Brown told Rep. Mel Price (D-IL), chairman of the subcommittee and a fool for all things nuclear, that "in view of the high cost of developing examples of modern technology and applying them to make a useful military system, it is incumbent upon us to evaluate very carefully the nuclear-powered ramjet vehicle, as to its technical development problems, its costs, and its schedules, as well as its military value in relation to other technologies or systems currently underway or being proposed."

The JCAE didn't want to hear Brown's objections. James Ramey, the joint committee's executive director (and later a member of the AEC), pushed Brown to move aggressively on Pluto. Brown bobbed and weaved. "Just how these will in the end compete with other possible ways of doing the same things, or how well it may compete with other ways of doing different things, we cannot say yet," he responded. The Kennedy administration proposed a low-ball $7 million in the Air Force budget for Pluto for fiscal 1962 (on top of $30 million for Pluto in the AEC account), far below the $24 million the Air Force wanted. The Air Force brass did a typical end-run around the civilian Pentagon, getting the full amount for Pluto from the joint committee and the appropriations committee, which often did or outdid the joint committee's bidding on matters nuclear.

But Brown and the civilians had sown seeds of doubt; even military planners outside of Livermore and Las Vegas were having some heartburn about Pluto. One problem was that the more conventional chemically-powered ballistic missiles could be launched safely from U.S. territory and get to their targets faster than Pluto. The missile, which gave itself away to adversaries from the moment of flight, was following the same fate as the atomic bomber. It was turning out to be less capable than first envisioned. Wags were saying that SLAM stood for "Slow, Low and Messy."

Then there was the daunting issue of how to test fly the weapon, which had the potential, as historian Herken put it, to become "a flying Chernobyl." The missile could not be flown over friendly territory. A crash landing could spread deadly radiation over thirty square miles.

So the scientists concocted a preposterous plan. They would fly the glowing bird in slow figure-eight paths over the South Pacific near a Wake Island takeoff. When the missile ran out of fuel, it would crash with a splash into the ocean. Herken quipped, "Pluto had begun to look like something only Goofy could love."

Project Pluto crash landed in July 1964, the start of a new federal fiscal year, abandoned by the Pentagon as a missile without a mission. The *New York Times* reported in early June, "The House Appropriations Committee dealt an apparent death blow today to Pluto, the low-flying, atomic-powered missile that has been having political trouble getting off the ground.

"The committee, in reporting out the Atomic Energy Commission budget, cut off development funds for the project. It proposed that the missile be 'mothballed' until the Pentagon could decide whether it had a requirement for the weapon."

10. Flatulence in Space

Jules Verne, the first practitioner of science fiction (although some might award that honor to the Bible), sent men to the moon in 1865. In 1898, H.G. Wells, who perfected the non-Biblical genre, sent Martians to the Earth in a conquering mood. Neither Verne nor Wells bothered with rocket propulsion. After all, Robert Goddard didn't invent what became the modern rocket until 1914, although he was pondering rocketry and space travel in 1902.[53]

While rockets are internal combustion engines, Verne and Wells both propelled their space travelers with forces outside of the space craft: explosions. Both Verne and Wells used artillery technology—chemical explosions pushing a payload (a cannon ball or a shell)—to move their creatures around the solar system. For a variety of reasons, this was a wildly impractical way to thrust heavy loads out of Earth's gravity and into space.

But the ideas of Verne and Wells lingered in the back of the minds of a trio of creative mathematicians and physicists, who would resurrect them the 1950s and early 1960s as Project Orion. Orion, which won some support from the AEC, the Air Force, and NASA, but was never embraced with the fervor of projects with a clear military future, aimed to use many small nuclear bombs exploding outside of their spacecraft to create gas plasmas shoving men and material out of the thrall of the Earth and beyond the planet's confining gravity.

Freeman Dyson, one of Orion's principals, described it some fifty years later, in his compilation, *Disturbing the Universe*. Dyson wrote,

---

53 Goddard was inspired by Verne and Wells.

Orion is a project to design a vehicle which would be propelled through space by repeated nuclear explosions occurring at a distance behind it. The vehicle may be either manned or unmanned; it carries a large supply of bombs, and machinery for throwing them out at the right place and time for efficient propulsion; it carries shock absorbers to protect machinery and crew from destructive jolts, and sufficient shielding to protect against heat and radiation. The vehicle has, of course, never been built. The project in its seven years of existence was confined to physics experiments, engineering tests of components, design studies, and theory. The total cost of the project was $10 million, spread over seven years, and the end result was a rather firm technical basis for believing that vehicles of this type could be developed, tested, and flown.

This improbable project—conceptually using atomic farts to blast manned capsules around the universe—drew a lot of enthusiasm in the U.S. nuclear flight enterprise for only a short period. Orion eventually died on the road to Jackass Flats, where the concepts that had been tested with chemical explosives in Southern California were scheduled to get a baptism of radiation. The project expired before the scientists could mount the atomic tests in part because it lacked a dogged government sponsor. During Orion's brief ascendancy, it drew some "gee whiz" reactions from President Kennedy. But in the end, the project succumbed not to its technical merits, but to Kennedy's success in working with the Soviet Union to limit nuclear testing.

Most of the adventures in atomic energy during the postwar period of irrational enthusiasm were the work of engineers. Orion was different. The three geniuses behind Orion—and there can be no doubt of their genius— were mathematicians Stanislaw Ulam and Freeman Dyson, and physicist Theodore Taylor. They worked together at a remarkable private company (although as a government contractor) General Atomics, located in one of the most beautiful sites in the world, overlooking the Pacific Ocean near San Diego. They were brought together and harnessed to space flight by another groundbreaking physicist and remarkable entrepreneur of modern physics, Frederic de Hoffmann.

Ulam was a Polish Jew, born in 1909 in what is now the Ukrainian city of Lvov to a wealthy family. His bent for math became clear early in life: he taught himself calculus at sixteen. He earned a PhD in mathematics from the Lvov Polytechnic Institute in 1933. Invited by Hungarian computing pioneer John von Neumann to visit the Institute for Advanced Studies in Princeton, New Jersey, in 1935, Ulam met Harvard mathematician George D. Birkhoff. Birkhoff invited him to join the newly-created Society of Fellows at Harvard. In 1936, Ulam began teaching in Boston, commuting in the summers back to Poland until World War II began in the fall of 1939.

The University of Wisconsin named Ulam an assistant professor in 1940, and he became a U.S. citizen in 1943, when von Neumann asked Ulam if he would be interested in some work, requiring citizenship, related to the war. Ulam described his introduction to the Manhattan Project: "[Von Neumann] discussed with me some mathematics, some interesting physics, and the importance of this work. And that was Los Alamos at the very start. A few months later I came with my wife...arriving for the first time in a very strange place." Thirty-four years old, Ulam was one of the older scientists working at the secret site in the remote New Mexican high desert.

At Los Alamos, Ulam linked up with Edward Teller to work on the many difficult problems involved in getting hydrogen atoms to fuse, producing an explosion that dwarfed even the terrible power of the atomic bomb. In the minds of several historians, Ulam, working with his former Wisconsin colleague Cornelius Everett, bore more responsibility for the H-bomb than Teller.

In 1946, while recuperating from brain surgery, Ulam came up with the all-important Monte Carlo method of statistically estimating the paths of neutrons in an atomic reaction. Ulam played a lot of solitaire while recovering from the surgery and noticed that he could predict the outcome of his games in advance by a few clues and patterns early in a game. "It occurred to me then that this could be equally true of all processes involving branching events," Ulam wrote years later. "At each stage of the process, there are many possibilities determining the fate of the neutron. It can scatter at one angle, change its velocity, be absorbed, or produce more neutrons by a fission of the target nucleus, and so on."

Based on this insight, Ulam came up with a way, using the primitive computers of the day (one of von Neumann's contributions), to make predictions of neutron behavior necessary for the hydrogen bomb calculations. Ulam and Everett analyzed Teller's design for the Super bomb. Ulam used his statistical methods and fertile mind to demonstrate conclusively that the design Teller doggedly held onto as a way to get the compression necessary to ignite and fuse hydrogen into helium could not work. Ulam then came up with a design using a fission explosion to compress the hydrogen isotope tritium to the point where the atoms would fuse, releasing an explosion of energy of mind-boggling force. Many histories and texts refer to the H-bomb as the "Teller-Ulam device."

Ironically, a driving force behind Ulam's work appears to have been his dislike of Teller, an animus Teller reciprocated. J. Carson Mark, a theoretical physicist who oversaw their work at Los Alamos, said, "Ulam used to make witty, pointed, scornful, shamefully disreputable remarks about Teller when Teller wasn't there. Once in a while his feelings about Teller couldn't have escaped Edward's notice. Edward reciprocated those feelings generously, so each was talking down the other, and that went on for years."

During the pioneering work on fusion, Ulam was already thinking about using nuclear explosions to send unmanned aircraft into space. He began discussing the idea with fellow mathematician Everett as early as 1946. Ulam came up with some preliminary calculations in 1947, in a document that is still classified. Mahaffey described the period as "the awkward time between the triumph of the atomic bomb and the push for the hydrogen bomb," adding that "Ulam's mind wandered off into the proposals for space exploration."

In August 1955, after the H-bomb test (code named Ivy Mike) on the Eniwetok atoll in the far South Pacific's Marshall Islands destroyed the Elugelab islet, Ulam and Everett wrote a detailed paper on the concept of nuclear explosions to propel missiles into space. In that document, "On a Method of Propulsion of Projectiles by Means of External Nuclear Explosion," Everett and Ulam outlined "the use of a series of expendable reactors (fission bombs) ejected and detonated at a considerable distance from the vehicle, which liberate the required energy in an external 'motor' consisting essentially of empty space. The critical question about such a method concerns its ability to draw on the real reserves of nuclear power liberated at bomb temperatures without smashing or melting the vehicle."

Earlier in 1955, Ted Taylor, the premier fission bomb designer at Los Alamos, worked with physicist and military officer Lew Allen, assigned by the Air Force to the New Mexico weapons lab, on a way to produce greater amounts of tritium, the rare hydrogen isotope needed in ever smaller and more powerful bombs. One experiment hung some iron and graphite balls from a bomb tower. After the bomb exploded, the scientists discovered that the spheres, predictably dubbed "Lew Allen's balls," traveled farther than could be explained simply by the force of the explosion and were not destroyed by the blast.

Taylor, who had emerged as a major figure at Los Alamos, began thinking seriously about Ulam's ideas for using explosions to propel space vehicles after he saw the evidence from Lew Allen's balls. "Being able to preserve things that were within twenty feet from the center of the explosion of tens of kilotons was a big surprise to a lot of people," he said. Freeman Dyson, who signed onto the Orion project four years later, said, "These experiments helped us to persuade people that the idea of an Orion ship surviving inside a succession of fireballs was not absurd."

Ulam's original name for his propulsion compulsion was Helios (named after the Greek god of the sun). In 1958, Ted Taylor, who was then heading the project at the General Dynamics subsidiary General Atomics, renamed it Orion, which he said he picked "out of the sky." Orion, the hunter, is one of the most prominent constellations in the sky. But Taylor's never lit up the sky. It never got off the ground.

Taylor's term at GA linked the theoretical work of Ulam, the peculiarity of Lew Allen's balls, and his own conceptual ideas to the practical attempt to turn imagination into reality. Taylor proved to be a most practical genius in turning out ever smaller and more efficient nuclear weapons during his days at Los Alamos; he was also a most intuitive thinker, a designer who left his intellectual DNA across many of the threads of early nuclear research and development, including Project Orion. One historian described him as "halfway between an inventor and a scientist." Carson Mark, head of theoretical studies at Los Alamos and Taylor's boss, described Taylor's approach to physics as "qualitative," as compared to the conventional, quantitative mindset that characterized most of the Los Alamos researchers.

Born in Mexico City in 1925, the son of a YMCA official, Taylor quickly demonstrated a facility for blowing things up, and a deep yearning, combined with a scent of fear, for space travel. As writer John McPhee described it in the fine 1973 book, *The Curve of Binding Energy*, Taylor as a boy "began to have recurrent dreams that would apparently last his lifetime, for he still has them, of worlds, planets, discs filling half his field of vision, filling all his nerves with terror. And yet he could not imagine anything more exciting than having travelled to and being about to land on Mars. He wanted to go there desperately."

After earning an undergraduate degree in physics from Cal Tech in 1945, Taylor entered the Navy, which, typical of the talent scouts of the U.S. military, assigned him to a ship transporting soldiers and sailors back to the United States from the war in the Pacific. The service ignored his entreaties for a slot in the atomic bomb program.

After mustering out of the Navy in 1946, Taylor enrolled in a PhD program in physics at the University of California–Berkeley. Uncomfortable with the pace and challenges of academic life, Taylor flunked out. Years later, he told McPhee, "Thermodynamics was dull. Sometimes I think I am incapable of understanding something I am not interested in. I studied it but did not learn. A lot of physics was a mystery to me, and still is."

While the mind-numbing intricacies of practical physics eluded Taylor, he thrived on theoretical courses. Mathematician Robert Serber, who became Taylor's mentor, recommended him for a slot at Los Alamos where Taylor's capacity, as McPhee put it, as a "conceiver of things" flourished. Moving to New Mexico in 1949, Taylor rubbed and exchanged sharp intellectual elbows with giants such as Ulam, Oppenheimer, and Teller and formed a lasting friendship with Allen, who went on to become Air Force chief of staff, the highest-ranking uniformed officer in the service.

In 1953, Taylor took a leave of absence from Los Alamos to complete a PhD under Hans Bethe at Cornell University in Ithaca, New York. There, he met Freeman Dyson, who had studied under Bethe (although Dyson never earned a doctorate) and was teaching. Degree in hand, Taylor returned to New Mexico and remained at Los Alamos until 1956.

Of Taylor's career at Los Alamos, Dyson wrote in a memorial essay, "He was famous in the community of bomb experts as the most creative and imaginative of the designers. His bomb designs were the smallest, the most elegant and the most efficient. He was able to draw his designs free-hand, without elaborate calculations. When they were built and tested, they worked." But Taylor was feeling a bit guilty about his bomb-making prowess, looking for a way to harness his skills to peaceful uses of atomic energy.

Freddy de Hoffmann, a Viennese-born physicist raised in Prague, Czechoslovakia, had immigrated to the United States in 1941, at the age of seventeen. De Hoffman was plucked out of Harvard in 1944, before his graduation, to work in Los Alamos, where he became chief assistant to Edward Teller from 1949 to 1951. At Los Alamos, de Hoffmann knew, worked with, and admired both Ulam and Taylor. De Hoffmann left Los Alamos for two years to complete his undergraduate degree and a masters and doctorate in physics at Harvard. During that period, he worked with Cornell's Bethe, another Los Alamos veteran, as well as Taylor's thesis advisor and Dyson's mentor.

De Hoffmann became one of two Americans (he became a citizen in 1946) to join the international secretariat staffing the 1955 Geneva arms control meeting, where he determined to devote himself to the peaceful uses of atomic energy.[54] At the same time, General Dynamics, the nation's largest defense contractor (among other ventures, they owned the Electric Boat submarine manufacturer in Groton, Connecticut) was looking to capitalize on the business promise of nuclear power. GD Chairman John Jay Hopkins went to Los Alamos and asked Teller whom he would pick to run such a venture, no doubt hoping Teller would nominate himself. Teller, knowing de Hoffman's skills as an entrepreneur and salesman, told Hopkins to hire de Hoffmann.

Hopkins in 1955 gave de Hoffman a line of credit of $10 million and told him to start what became General Atomics, headquartered in La Jolla, California. The first scientist de Hoffman hired was Taylor. In the summer of 1956, while getting GA underway, de Hoffmann invited a spectacular

---

54 De Hoffmann eventually switched interests to biomedicine and, in 1970, joined the Salk Institute, founded by polio vaccine discoverer, Jonas Salk. As president of the institute of La Jolla, California for eighteen years, de Hoffmann presided over a period of enormous growth and great accomplishments. He died in 1989 at the age of sixty-five from AIDS, which he acquired from a blood transfusion during open heart surgery in 1984.

group of luminaries from atomic physics and mathematics for a summer institute held in a renovated schoolhouse on the California beach near Point Loma. Among those who attended were Teller, Bethe, and Dyson. "Our primary job," he said, "was to find out whether there was any specific type of reactor that looked promising as a commercial venture for General Atomic to build and sell."

Teller suggested development of an inherently safe reactor, one that even clumsy graduate students (which Dyson admitted characterized his career as an experimentalist) couldn't bring to the point of danger. Teller knew that few institutions would be willing to experiment with reactors unless they were supremely safe. Their work resulted in TRIGA (Training, Research, and Isotopes, General Atomics), without a doubt the most successful nuclear reactor of all time, and eventually a prodigious moneymaker for GA,[55] after debuting on the market in 1959.

After the summer of sun and physics on the California beach, Dyson returned to his home at Princeton's Institute for Advanced Studies in New Jersey. While Dyson was pondering questions such as the basis of quantum energy, the Soviet Union put Sputnik into orbit, and Taylor, in Los Alamos, was focusing on using bombs to push large loads into and through space. On the same October 1957 day that the USSR announced its space triumph to the world, Taylor issued a technical paper titled "Note on the Possibility of Nuclear Propulsion of a Very Large Vehicle at Greater than Earth Escape Velocity," and dubbed the craft Orion. He quickly began lobbying the Eisenhower administration for funding, finding a sympathetic ear in the Air Force's special weapons center at Kirtland Air Force Base in New Mexico, where Lew Allen was the science advisor.

Dyson recalled that in 1958, "Freddy de Hoffmann passed through Princeton and told me the latest news of the operational trials of the prototype TRIGA. 'By the way,' he said, 'Ted Taylor has a crazy idea for a nuclear spaceship, and he wants you to come out to San Diego and look at it.' I went." Dyson added that the backstory was Sputnik, and the U.S. plans for a major response, using chemical rockets to land men on the moon in

---

55 TRIGA is the most widely-used non-power reactor in the world. GA over the years has sold sixty-six TRIGA machines, according to GA, which continues to earn profits from the research and isotope production reactors.

ten years, at a cost of $10 billion or so. "Ted was interested in going into space," Dyson recalled, "but was repelled by the billion-dollar style of the big government organization."

The addition of the cerebral and lyrical British mathematician Dyson was the final ingredient to what became the Orion team, a group which had a fine time blowing up models in California with conventional explosives and dreaming of trips to Mars and Saturn. Dyson agreed to join GA for the 1958–1959 academic year to refine the concepts behind Orion and begin to test those concepts empirically. Dyson said, "We intended to build a spaceship which would be simple, rugged, and capable of carrying large payloads cheaply all over the solar system. Our slogan for the project was 'Saturn by 1970s.'"

Dyson also shared Taylor's passion to use the tools of wartime destruction for peaceful purposes. "We have for the first time imagined a way to use the huge stockpile of our bombs for better purpose than for murdering people," he wrote. "Our purpose, and our belief, is that the bombs which killed and maimed at Hiroshima and Nagasaki shall one day open the skies to man."

In 1958 Taylor made a proposal to the Pentagon's Advanced Research Projects Agency (ARPA) for a mammoth four-thousand-ton spacecraft, to be powered by some twenty-six hundred small atomic bombs dropped out of the rear end and exploded, pushing the craft up from the Earth and into space. The Orion craft would carry a sixteen-hundred-ton payload, an order of magnitude beyond anything that was then achievable with conventional chemical rockets. He estimated it would be ready to fly by 1963–1964, at a cost of about $500 million, an estimate surely as wildly plucked out of the sky as the name Orion itself.

Mahaffey described the technology of Orion: a small nuclear device would be shot out the back of the craft through a tube, using compressed air. It would ignite two hundred feet behind the vehicle, "putting its back end just out of reach of the fireball from the explosion." A thick steel pusher plate would catch the explosion and push the craft ahead, while the compartment holding the crew would be protected by giant shock absorbers. "Each explosion [would] add 30 miles per hour to the forward speed." If the bombs detonated every three seconds, the craft would reach 3,000 mph in five minutes, making long-distance space travel possible.

Taylor first took the project to the White House for blessing. Eisenhower's science advisor, George Kistiakowski, considered including the project as part of the Atoms for Peace program, but ultimately turned it down.[56] Stan Ulam in January 1958 testified in support of the Orion project at a Joint Atomic Energy Committee hearing in Washington. "It is almost like Jules Verne's idea of shooting a rocket to the moon," Ulam told the atomic solons.

Despite his reluctance to push the project's military utility, Taylor (with help from the Air Force's Allen) later in the year closed a deal with ARPA for $1 million in fiscal year 1959 (beginning July 1, 1958). In July 1958, ARPA issued a press release noting the $1 million allocation of FY59 funds to GA for a "new concept of propulsion employing controlled nuclear explosions."

The ARPA announcement provided evidence of a problem that would plague, and ultimately kill, Orion. It wasn't clear whether the spaceship propelled by atomic flatulence was an example of atoms for peace or for war. Over the few years of its perilous bureaucratic life, Orion won funding both as a military project and a civilian initiative, but never won the love of either the military or the civilian space voyagers.

Orion came to life as the United States was flailing in response to the Russian flight of Sputnik. At the point of policy panic, the United States appeared ready to place bets on every number on the roulette wheel to space. George Dyson, Freeman's son and later chronicler of the Orion project, described the timing as "a narrow window of opportunity between the launch of Sputnik and the commitment of the United States to an exclusively chemical approach to space." But GA was unable in 1958 to convince either the Air Force or the newly-created NASA of the ultimate utility of Orion, winning only the small feasibility study from ARPA—which had a broad mandate to support research and development, whether the projects had military consequences or not.

With ARPA funding in hand and Dyson on the scene, Orion began its most productive period. By all accounts, it was great fun for all involved. The work consisted of making theoretical hypotheses and mathematical

---

56 In 1959, the AEC patented the idea of a bomb-powered spacecraft, crediting Ulam and Everett.

prognostications and then testing them by using copious amounts of conventional explosives (usually the very potent C-4) to see whether hypothesis made it into the realm of reality. Often it did, leading the physicists and engineers, mostly recovering Manhattan Project weaponeers, to earnestly believe that what worked at bench scale with chemistry would succeed in the radioactive crucible at Jackass Flats. Dyson described the eighteen months he spent in the sun and surf at La Jolla as one of the happiest times of his life.

In February 1959, the Orion band made the first of a bold series of tests on a three-foot diameter model of the craft, with results that emboldened the designers to believe that success at full scale—some thirty times the size of the model—was likely. It whetted their scientific appetites for more—and more realistic—tests. But technically, these tests were illegal. The ARPA contract contained unequivocal language forbidding modeling. The GA management took its case to ARPA, which amended the contract in mid-1959 to permit a full flight test of the three-foot Orion model. On November 14, 1959, looking remarkably like artists conceptions of Jules Verne's ballistic space craft, the mini-Orion lifted off from Point Loma, and rose to 185 feet, where a parachute deployed and the missile floated back to Earth. That was the last time anything resembling Orion lifted off the ground.

While the model space ship was preparing for its short flight, and after, the Orion project was putting plenty of effort into the design of a full-scale, manned craft that would first require extensive testing at Jackass Flats. But project funding was running out, and GA had to scramble to keep the project intact.

Over its short history, Orion was a radioactive volleyball, bouncing among ARPA, NASA, and the Air Force. With the first ARPA round of funds disappearing, Taylor and Dyson persuaded the Pentagon to put up another four-hundred thousand dollars to keep the project alive while Washington sorted out who should own Orion. NASA, a logical candidate, wanted nothing to do with passing nuclear gas. The Air Force wasn't much interested. They had bigger fish—the A-plane and Pluto—in the atomic frying pan. Also, the flyboys couldn't perceive much of a military mission for Orion. ARPA dropped the project in 1960. Thanks to the support of

Allen, the Air Force stepped in rather reluctantly to fund Orion. NASA put up a small amount for the project in 1963.

While it stayed on life support until January 1965, Orion effectively went into eclipse in 1963, when the Kennedy administration got the Soviet Union to agree to a treaty banning atmospheric testing of nuclear bombs. Whatever the virtues of Orion, it fundamentally depended on exploding thousands of nuclear weapons, either on the ground or in space.

Ironically, Freeman Dyson, who had returned to Princeton's Institute of Advanced Studies in late 1959 but remained an Orion enthusiast, helped kill the program. Perceived as a hard-liner on nuclear weapons, Dyson became a major witness supporting the limited test ban treaty in Senate ratification hearings, testifying on behalf of the Federation of American Scientists and providing technical advice to the Kennedy administration in support of the test ban. "I met Ted Taylor in Washington" in the fall of 1963, Dyson later wrote, "and told him I had signed Orion's death warrant."

In a somewhat dyspeptic article in *Science* magazine in 1965, Dyson groused that the decision to kill Orion was "the first time in modern history that a major expansion of human technology has been suppressed for political reasons." Fifteen years later, he amended his views. "Sometimes I am asked by friends who shared the joys and sorrows of Orion whether I would revive the project if by some miracle the necessary funds were to become available," he said. "My answer is an emphatic no...I would not now wish to fly about in a ship that dumps radioactive debris upon the heads of the passengers in our other spaceship, Spaceship Earth." He was even more dismissive when quoted in a 2010 article in *Atlantic* magazine: "The starship was like an existence theorem in math. It was to prove you could do it. I never really believed in it."

The attempts of the atomic establishment to fly with nuclear power from the Earth into space proved to be failures. The ventures were either impossible (the bomber to nowhere), impractical (Rover and Orion), or intolerable (Pluto). But decades later, there were still plenty of cockeyed nuclear optimists around who somehow hoped to revive the flights of fancy.

In the meantime, the United States had also embarked upon another great adventure and feckless use of nuclear technology— not using nukes to fly, but to dig into the ground and accomplish grand feats of terrestrial engineering. Edward Teller, a man whose lust for nuclear technology never cooled, once again championed the cause—one of his greatest enthusiasms.

Eddie Teller's Exploding Ambitions

In *Tom Swift and His Atomic Earth Blaster* (1954), the scientifically precocious teen, after already inventing a nuclear powered airplane and an atomic submarine, turned his talents toward moving earth. Tom harnessed a nuclear pile for the motive power and heat to blast and melt rocks. Questioned whether the device could be used by nefarious foreigners as a weapon of war, Tom replied, "But the earth blaster is for peacetime use. It's not a weapon that could be used for fighting a war."

Or could it?

11. The Atomic Earth-Blaster

Edward Teller, father of the hydrogen bomb and inspiration for Terry Southern's wonderful screenplay *Dr. Strangelove or: How I Stopped Worrying and Learned to Love the Bomb*, had a revelation in the mid-1950s—a case of life imitating art (or at least life imitating Tom Swift Jr.). Throughout his lifetime, Teller saw H-bombs as instruments of peace: first to rearrange inconvenient aspects of geography and geology, later to erect a space-based nuclear shield against incoming Soviet ballistic warheads, and finally to blast incoming asteroids off paths that would crash them into Earth.

Teller early conceived of using H-bombs to rearrange the shape of the planet—for peaceful uses, of course. Thus, they called it the Plowshare program. Teller had met with Eisenhower in 1955, offering the elderly president visions of the peaceful atom, including earthmoving applications. These ideas coincided with Ike's major international policy objective, Atoms for Peace, launched two years earlier.

Over the next two decades, the United States would repeatedly attempt major atomic earthmoving projects, shooting off a series of feckless and often dirty nuclear explosions in an orchestrated program of what some have called "geographical engineering." In the end, nothing physical was accomplished except a couple of obscure holes in the ground.

But among its legacies—validating the theory of unintended consequences—the Plowshare endeavor ultimately contributed greatly to the growing environmental and anti-nuclear movements, adding additional dimension to the problem of radioactive fallout and damaging the credibility of government institutions, particularly the Atomic Energy Commission and its congressional overseers. The program, no doubt to Edward Teller's dismay, also helped propel the United States and the Soviet Union into a lasting series of atomic test ban treaties. The fallout from Plowshare, commented one historian of the program, was "first radiological, then political."

The civilian atomic energy project Teller pushed through Congress and the Atomic Energy Commission gained the Biblical name Plowshare from young physicist Harold Brown. The name came from the admonition in the Old Testament Book of Isaiah (2:4): "They will beat their swords into plowshares and their spears into pruning hooks. Nation will not take up sword against nation, nor will they train for war anymore."[57]

At the time of his earthmoving insight, Teller was working at Ernest O. Lawrence's University of California Radiation Laboratory at Livermore. Teller had persuaded the AEC to create Livermore as a second national atomic weapons laboratory and a competitor to the Los Alamos National Laboratory in New Mexico. Los Alamos was the fruit of his rival and nemesis, J. Robert Oppenheimer. While the government was initially indifferent to Teller's ambitions, the plan for a second national nuclear weapons laboratory took off after the Soviet Union exploded its first atomic bomb in 1949.

A brilliant Hungarian physicist who immigrated to the United States as Germany conquered Europe and savaged Jews, Teller felt stifled at Los Alamos in the Manhattan Project. He pushed for the development of a thermonuclear (hydrogen fusion) bomb to succeed the fission bomb that Oppenheimer developed. Oppenheimer opposed the development of the fusion Super bomb, as did many Los Alamos scientists.

But Teller ultimately prevailed. The H-bomb was a sensational success, vaporizing a Pacific atoll and spreading inadvertent nuclear fallout over much of the world. Los Alamos had grown too remote and too confining for Teller's atomic ambitions. Lawrence's Livermore lab was a broader canvas for Teller's atomic arts.

---

57 Coincidentally, when he was sworn in as AEC chairman in 1953 by President Eisenhower, Lewis Strauss took the oath with his hand placed on a Bible turned to a passage in the Old Testament book of Micah (4:3): "And he shall judge between many peoples, and shall decide concerning mighty nations afar off; and they shall beat their swords into plowshares, and their spears into pruning hooks; nations shall not lift up sword against nation, neither shall they learn war anymore."

Once he and the legendary Lawrence[58] convinced the government to establish a second nuclear weapons lab at the Livermore radiation laboratory in 1952, thirty-six-year-old Teller turned to projects that would confound his former colleagues at Los Alamos. Livermore would become the challenger to all the Los Alamos bomb projects, and the originator of new projects to use nuclear bombs, particularly Teller's H-bombs, in the service not only of military destruction but of nominally peaceful, civilian purposes.

Nuclear energy was a game for young men, particularly at the California offshoot of Los Alamos. Teller was born in 1912. Herbert York, the first Livermore lab director, who later became a key science advisor to presidents Eisenhower and Kennedy, was thirty-two at the time Livermore was created. Brown, who supported Teller's visions of peaceful nuclear explosion and succeeded Teller as Livermore director in 1960 (and became defense secretary under President Jimmy Carter), was born in 1927.

Lawrence, the guiding hand behind the laboratory, the éminence grise, was still only fifty-one when the AEC created the second weapons lab, and had been a star of international physics for over twenty years. While most of the bright lights at Livermore were theoretical physicists such as Teller, Lawrence was an experimentalist, a machine builder who valued tangible data that validated theory. During its early years, the Livermore laboratory not only radiated nuclear energy, but the personal energy of its young and ambitious staff.

In 1956, Livermore floated the notion of using nuclear explosives to clear and move dirt and rocks, when the AEC cautiously authorized a secret symposium on the subject. Brown, working with his boss Teller, organized the event, which took place February 6–8, 1957, at Livermore in the California desert. The event brought together scientists from across the widespread U.S. nuclear weapons complex to contemplate a new mission for their weapons.

Teller kicked off the conference with a typically hyperbolic and creative approach to using nuclear explosions. He literally proposed to shoot the

---

58 Lawrence won a Nobel Prize in physics in 1939, for his invention ten years earlier of the cyclotron, popularly called an "atom smasher," a machine that was essential to understanding the nature and composition of the atom.

moon. "One will probably not long resist the temptation to shoot at the moon," said Teller. "The device might be set off relatively close to the moon and one would then look for the fluorescence coming off the lunar surface, or one might actually shoot right at the moon, try to observe what kind of disturbance it might cause."

While Teller was focused heavenward, the symposium took a somewhat more practical down-to-earth turn. There was considerable enthusiasm for nuclear landscaping, particularly directed at digging a new, sea level Panama Canal. In concluding remarks, Brown described the future of the global earthmoving program, which he saw as "a group of some number, such as ten or twenty people, thinking about it for six months or a year, picking out the good ideas, and working them out in some detail." It didn't take that long. By July, the Plowshare program was officially a Livermore mission, under the titular leadership of the AEC's Division of Military Applications in Washington, working through the AEC's San Francisco office.

The Plowshare program got a major boost in September 1957, when the AEC set off a 1.7-kiloton blast nineteen hundred feet inside a mesa at the agency's Nevada Test Site. Code named Ranier, the test demonstrated that underground explosions were feasible. The test resulted in no radioactive fallout. The AEC was attempting to understand the feasibility of underground testing and detection, but AEC Commissioner Willard Libby, the only scientist on the commission, saw it as justifying a program of peaceful explosions, as outlined at the Livermore conference seven months earlier. Libby told an Amherst College audience in October that Ranier validated his notions of use of atomic bombs for "non-military or peaceful applications of nuclear explosions, and the possibility that much of the testing can in the future be done under conditions of no radioactive fallout."

Teller was truly interested in ways to use the power of the atom to help humanity. He was also a fervent anti-communist who wanted to forestall the advance of the Soviet Union by any means. Ultimately, many scholars concluded that Teller's plan to rearrange the landscape by "geographical engineering" with atom bombs was as much about continuing testing bombs in the face of internationally-agreed-upon treaties limiting nuclear weapons as a desire to accomplish civilian ends. Facing the horror of a nuclear war between the United States and the Soviets, and the fallout from atmospheric tests by both nuclear powers, the concept of limiting testing, and ultimately limiting the arms themselves, was in the air. Teller found that atmosphere threatening.

Freeman Dyson visited Livermore in 1958 and found it "wildly exciting." He wrote home: "A lot of talk at Livermore was about cheating the test ban. We found a lot of ways to cheat which would be quite impossible for any instruments to detect. The point of this is not that the Livermore people themselves intended to cheat, but was that they are convinced the Russians can cheat as much as they want any time they want, without being found out."

The international situation was an important component throughout the AEC's program planning. The commission was both the nation's civilian nuclear agency, and—more importantly—the designer, maker, and often de facto policymaker of the nation's nuclear weapons program. At the 1957 suggestion of the Soviet Union, the world's nuclear powers—the United States, Britain, and the USSR— agreed in October 1958[59] to a voluntary moratorium on future above-ground nuclear bomb tests, while they discussed a permanent ban. Eisenhower, at the urging of the AEC, defined the limits of the moratorium to allow explosions aimed at peaceful uses. The voluntary test ban expired in 1961, as the world was negotiating for a formal follow-up: the 1963 Limited Test Ban Treaty.

In an October 16, 1958, secret letter to then–AEC chairman John McCone, when the Eisenhower administration put in place its portion of the voluntary international moratorium on atmospheric testing, Teller wrote that much of Livermore's activities would "depend to a considerable extent on the precise terms of the moratorium. If we assume that only such tests will be banned which actually can be policed, then a considerable amount of testing can accompany the further development of weapons." Livermore, Teller wrote, will "increase greatly our efforts in the non-military uses of nuclear explosives…We expect that the Plowshare program will account for 4 percent of the laboratory's effort in the next period."

During the two-year interregnum between the voluntary ban and the new limited treaty, there was much discussion about the distinction

---

59 France became the fourth nuclear weapons state in 1960, exploding a 70-kiloton device in the Algerian desert.

between above-ground and underground testing, as well as debate about what constituted "peaceful" tests. AEC commissioner Willard Libby told a university audience, before the voluntary moratorium, "We must proceed with the non-military applications of nuclear explosions under any conditions! Let us all hope these will be conditions of real disarmament fully implemented and inspected."

The distinctions in the arms control discussions often turned out to be blurred. If a "peaceful" explosion developed data useful for military planners—whether that was part of its stated mission or not—what was it? If a bomb placed underground actually broke the surface, either by design or accident, was it an underground or above-ground blast? If fallout from a peaceful experiment crossed international borders, was that a treaty violation? Debate swirled around these and other questions.

The concept of turning weapons into peaceful engineering tools had a clear propaganda purpose, as well as constituting an attempt to harness the atom to peaceful purposes. A major aim of the program, openly stated, was to allay fears of the atom and atomic testing among the public. In closing the Livermore conference, Brown said, "In the past twelve years, all kinds of phobic public reactions have been built about nuclear bombs," implying that blasting useful harbors and canals with the atomic bombs could help change that attitude. Eisenhower's Atomic Energy Commission chairman, Lewis Strauss, was even more forthcoming than Brown about the public relations mission of Plowshare. In 1958, Strauss said Plowshare's goal was to "highlight the peaceful application of nuclear explosive devices and thereby create a climate of world opinion that is more favorable to weapons development and tests."

Teller had great ambitions for Plowshare. He wanted to use nuclear bombs to dig out a new harbor in rural Alaska; blast a sea level ditch across the Isthmus of Panama; open a rail route through the Mojave mountains from California to the east; release oil from shale and sand and natural gas from so-called tight deposits in the Rockies; make steam to generate electricity; and mine caverns for storage of natural gas in Pennsylvania. His vision was broad and breathtaking, although, like much of his endeavors following the H-bomb, it proved too far-reaching. So devoted were Teller and his California crew to lighting off nuclear explosives to rearrange the landscape that they earned the tongue-in-cheek sobriquet the Firecracker Boys from Alaska historian Dan O'Neill.[60]

The most visible result over more than a decade of planned and realized explosions was an enormous, useless, and radioactive crater in the desert of the Nevada Test Site south of Las Vegas, along with numerous violations of international test ban treaties, before the program officially went off the government's books in 1974. No canals, no harbors, no new sources of fossil fuels, no new approaches to making electricity. Between 1957 and 1974, the Plowshare program detonated twenty-nine nuclear bombs at the Nevada test site and another six in New Mexico and Colorado. A planned explosion in central Pennsylvania aimed at creating a cavern for storing natural gas never got off the ground. While the AEC and Livermore downplayed the environmental results of the Plowshare explosions, several resulted in major radiation releases to the atmosphere, causing closure of highways, wash-downs of vehicles, and contamination of milk produced by local cows. During that time span, the AEC spent some $770 million (in 1996 dollars) on the Plowshare program, employing a couple hundred full-time workers annually at Livermore.

---

60 O'Neill took that moniker as the title for his fine 1994 book on Plowshare's Chariot program, *The Firecracker Boys*.

12. Chariot Swings Down to Alaska

Teller's first Plowshare target was to rework the landscape in remote, rural Alaska. The bull's-eye was a desolate site in northwest Alaska on the Chukchi Sea, where the Atomic Energy Commission proposed to carve out a new ocean harbor with a series of H-bomb blasts. Livermore called it Chariot; many of the Plowshare projects, for unknown reasons, had names related to methods of ground transportation: Sedan, Buggy, Schooner, Sulky, Palanquin, Ketch, and Cabriolet.

Teller wanted to use the Alaska project to demonstrate the atomic earthmoving technology and its potential for good. Chariot was, in short, an initial marketing effort. At a press conference in 1960 in Alaska, Teller said, "If your mountain is not in the right place, drop us a card."

Why Teller chose the specific site or the name of the Alaskan project is lost in the mists of history. Still, Teller and his crew clearly did not arbitrarily choose the remote site. Teller and AEC chairman Strauss had decided as a political matter that it would be impossible to sell the technology outside the United States (Panama, for instance) unless the United States was able to first demonstrate its safety and efficacy at home. The AEC was also wary of spreading atomic fallout over an unsuspecting population, as had occurred with the first tests of the H-bomb in the South Pacific.

Alaska, remote from the lower forty-eight states, with few inhabitants, and not yet a state itself, seemed ideal for the first demonstration. Alaska consisted almost entirely of federally-controlled land; the territory had little political clout in the larger nation. In early 1958, the AEC's San Francisco office asked the U.S. Geological Survey, part of the Department of the Interior, to analyze the prospects for blasting out a harbor on the Alaskan coast between Nome and Point Barrow—an enormous swath of sparsely-settled territory with the distance between the two settlements about three hundred miles through the air (with a sixteen hundred-mile coastline). Later the AEC clarified its USGS request, asking for a study of a twenty-mile coastal strip south of Cape Thompson, some twenty-six miles southeast of the native village of Point Hope. At the same time, Livermore hired a San Francisco contractor, E.J. Longyear Co., which did considerable work for the laboratory, to look at the mineral estate. Both the USGS and the private contractor issued positive reports on blasting an Alaska harbor. Neither actually traveled to Alaska to conduct the investigations.

The project timing was suspicious. As one of Teller's biographies notes, Teller publicly announced the Chariot project at a press conference in Juneau, Alaska, in July 1958 "while the Conference on Experts was meeting in Geneva to discuss an end to nuclear testing." The Livermore press conference was a surprise to the Alaskans, as Teller and his crew from Livermore landed in the territorial capital unannounced.

Teller's hastily-called Juneau press conference in 1958 unveiled his plans for a deep-water harbor at Cape Thompson, 160 miles across the sea from the furthest reaches of the Soviet Union. The area for the Chariot project constituted sixteen hundred square miles, considerably larger than the state of Rhode Island. But it was remote from any centers of political power, closer to Africa than to Washington, DC. Teller made enormous claims for the Chariot project. It would, he told the Alaskans, have great economic benefits for Alaska, including offering a new opportunity for commercial fishing (then and now the largest employer in the state). The new, man-made harbor would also allow development of significant coal deposits in that part of Alaska.

Teller said his bomb builders could precisely carve out the harbor, with no harmful radiation coming from the explosions. He said grandiloquently that the Livermore engineers could "dig a harbor in the shape of a polar bear if desired." Teller also said that in choosing the site in Alaska, "We looked at the whole world—almost the whole world." That was a bit of a lie, rescued only by the final clause. Teller himself had limited the search to sites in the United States.

Livermore characteristically moved quickly to ramp up the Chariot project, to some discomfort at the AEC in Washington. The *New York Times* reported in September 1958, in a short article on page six, date-lined Geneva: "The Atomic Energy Commission is about to drop plans to excavate a harbor in northern Alaska because nobody seems to want it. Dr. Willard Libby, a member of the commission, said today it had made a survey into the possibility of using nuclear explosions to carve out the harbor." Libby was at the international test ban negotiations when he made the statement. The only nuclear scientist on the commission at the time and a major designer of the gaseous diffusion project that liberated enough U235 from natural uranium to support the first atomic bomb, Libby was a strong supporter of the Plowshare program. But his remarks in Geneva reflected the doubts about Chariot within the commission.

Tension arose between the Plowshare program in California, the AEC in Washington, and some prominent Alaskans throughout the history of the Chariot blast. In 1959, new AEC chairman John McCone told the congressional Joint Committee on Atomic Energy, "We are seeking an alternative to the harbor in Alaska because, as I said to the committee once before, we couldn't find a customer for the harbor." Alaska's newly-installed Democratic senator[61] E.L. "Bob" Bartlett told the committee that "no one on the commission staff concerned with this believed for a minute private capital would, in the foreseeable future, invest money in this area merely because an artificial harbor had been created...I have now been completely disillusioned...For one, I hope the AEC does its blasting elsewhere."

But Teller was an accomplished politician and publicist. He kept the Alaska project alive for several years, despite a crumbling base of support among Alaska business interests and amid continuing doubts at the AEC. His force of personality and blinding brilliance overcame most political obstacles throughout his career. His problem, in the Chariot project and elsewhere, was that his imagination often exceeded physical and practical realities that ultimately undermined his dreams.

The Chariot announcement and press conference got little attention outside Alaska. The big national rollout came on June 5, 1959, in a *New York Times* front-page article written by science reporter Walter Sullivan. In an article with no attribution to named sources, and undoubtedly based on information from Teller and John Wolfe of the AEC's San Francisco office, where Wolfe was in charge of environmental analysis, Sullivan wrote, "Plans have recently been completed for an exhaustive study of a lonely

---

61 Alaska became a state on January 3, 1959.

stretch of Alaskan coast where five hydrogen bombs may be used to blast a harbor from the tundra.

"The plan is to fire one bomb near enough to the beach to carve out a channel. The four others would be grouped about three-quarters of a mile inland to produce the harbor basin. The site is the mouth of Ogotoruk Creek, near Cape Thompson, 175 miles across the Chukchi Sea from the Soviet Union." The AEC said it was hoping for the Chariot shots in "1960 or 1961."

Alaska historian Dan O'Neill observed, "Though these remarks played well with the newspaper editors and with many civic boosters, nearly every single material claim Edward Teller made in Alaska seems to have been untrue." That observation pertains not only to Teller, but to everything the Livermore national laboratory and the Atomic Energy Commission said about the Alaska project. Overstatements, misleading claims, obfuscations, and outright lies would later prove to be the standard operating procedure for the Plowshare program.

The *New York Times* article also drew attention from the Soviets. On June 14, according to the *Times*, the Soviet newspaper *Sovetsky Flot* commented, "The explosion of these bombs will be nothing but a camouflaged test of thermonuclear weapons. The reactionary forces of the United States are engaging in the impudent deception of people." The Soviets had an aggressive nuclear excavation program of their own in progress (in contrast to the case of the nuclear-powered airplane).

Predictably, the United States almost immediately began chanting the "here come the Russians" mantra to advance its atomic ambitions, although this time, there was substance to the claim. *Times* science writer Gladwin Hill wrote in April 1960, "The United States is moving slowly into a sort of hare-and-tortoise race with the Soviet Union in the field of massive underground explosions for engineering purposes." The article continued, "Plowshare scientists, whose headquarters are at the commission's laboratory in Livermore, California, are certain that Soviet knowledge has forged far ahead of American knowledge in this realm." Hill quoted the AEC's Gerald Johnson calling for the Soviets to reveal their data on atomic excavations. Johnson claimed that if they would release their data, "The Plowshare excavation program could be advanced by at least two years."

As Paul Josephson describes in *Red Atom*, the Soviets had a head start on "peaceful nuclear explosives," or PNEs, the jargon used during international negotiations to describe blowing up the landscape with nuclear bombs. By 1958, the Soviet Union had used nuclear explosions to dig diversion canals and clear forest.

Teller feared that the Soviets would exploit their lead in atomic landscaping in the race to win the hearts and minds of uncommitted nations. He wrote in a 1962 book, "The time may be near when the Russians will announce that they stand ready to help their friends with gigantic nuclear projects. The consequences of such aid would be an economic penetration a hundred times more extensive than those following the Soviet offer to help Egypt construct the Aswan dam." He also saw the competition between the United States and the Soviet Union as a propaganda war, with the Soviets demonstrating that they "certainly must be ahead of the United States in military applications. As a propaganda weapon, [the Soviet PNE program] could finish the work begun with the launching of Sputnik."

The AEC and Teller were back in Alaska in March 1960 for a press event to further tout Chariot, including the quip about moving mountains. At this press conference, the AEC unintentionally revealed that the agency had already prejudged the likely environmental consequences of the Chariot blasts. The *Times* reported that the AEC "said its committee on environmental studies for the project had set March or April as the preferred time of year for the detonation of blasts creating the new harbor." At this point, the AEC had made no ground-based environmental reviews in Alaska. The agency seemed oblivious to the fact that the area harbored a significant Eskimo population. Possibly, the agency didn't believe the project would harm humans, so the presence of a native population was irrelevant to the AEC project planners.

While the Chariot blast initially had wide support inside Alaska, that soon began to erode, particularly when biologists, geographers, and wildlife scientists at the University of Alaska started scrutinizing the technical claims and claiming a piece of the analytical action for themselves. The academic scientists quickly saw large holes in the AEC's hasty, remote environmental analysis. The skepticism of the local scientists threated public support for the project. In order to buy off the local academics, the AEC agreed to contract with the university to do baseline environmental studies at the proposed site. As that work progressed, it became increasingly clear that Livermore scientists knew very little about the flora and fauna at the site. When it came to assessments of the environmental effects of an H-bomb blast on remote Alaska, the atomic scientists at the AEC were faking it.

A cadre of University of Alaska scientists spread out over the Cape Thompson area in the summer of 1960, consistently finding facts and circumstances that belied the assertions by the AEC's chief environmental analyst John Wolfe. Among the findings was that the local Eskimo population depended on caribou hunted close to where the government proposed to set off the bulk of its atomic explosions. In turn, the caribou depended, particularly in winter, on lichens as fodder. The lichens were particularly well-evolved to take up the trace element strontium in their life cycle, passing the rare earth element on to the caribou, and hence to the native population. The fission bombs that trigger the thermonuclear blasts produced radioactive strontium, which would move up the food chain and could cause cancer in the Eskimo inhabitants.

Wolfe and the Atomic Energy Commission did everything they could to silence the Alaskan scientists, threatening to refuse to renew contracts for the scientists to continue their investigations unless they knuckled under to the AEC leadership. Some of the scientists also began to question whether they could continue to participate in a program that they believed was intellectually dishonest and seeking to hide problems with the site. On August 15, 1960, the AEC's Wolfe told a press conference at Point Hope, a tiny Eskimo village, which at thirty-five miles was the closest piece of civilization to the proposed blast site, that a fifteen-month, $2 million study found "no biological objections to the shooting on the basis of our investigations." Chariot would not damage the hunting and gathering lifestyle of the local Eskimos, Wolf asserted.

Wolfe's press conference produced considerable skepticism. Some Alaskans were beginning to sense a conflict between what the AEC was saying and what the environmental scientists on the ground were finding. Enter the Alaska Conservation Society, a small group based in Fairbanks that had recently won a battle to protect what is now known as the Arctic National Wildlife Refuge, some nineteen million acres in the Brooks Range in northern Alaska. The local group followed that success by targeting the Chariot project, seeking a delay until the wildlife and people in the region were protected, and calling for an independent environmental analysis.

The decision by the conservationists to target Chariot heartened the University of Alaska environmental scientists, who decided in early 1961 to go public with their concerns about the AEC studies—to which the Alaskan scientists had contributed all of the data. The three academics most concerned about the project, Don Foote, Les Viereck, and Bill Pruitt, wrote a detailed critique of the AEC's activities in supporting Chariot, for a thirty-page edition of the conservation society's *News Bulletin*. Pruitt summed up the biological issues, Viereck addressed botanical matters, and Foote, a fiery activist, contributed a highly-critical overview. The society printed one thousand copies of the March 1961 *News Bulletin* (a press run some four times the size of the organization's membership) and distributed it widely.

Pruitt sent a copy to a plant physiologist at Washington University in St. Louis named Barry Commoner, who was becoming known as an opponent of nuclear testing through the Greater St. Louis Citizens Committee for Nuclear Information (CNI). The CNI republished the Alaska report, drawing national attention to the Chariot plan. For Commoner, it was the beginning of a long national career as an environmental advocate and political gadfly.

As opposition mounted and Chariot began drawing fire from environmental groups across the country, the AEC repeatedly put back the date for the shot. A major problem, in addition to the green black eye the government was getting from the nascent environmental movement, was the advancing worldwide sentiment for nuclear arms control. In the August 1960 dispatch from Point Hope, the AEC was still insisting that there were no potentially harmful effects from the Chariot blast. But a *New York Times* reporter commented, "The commission has not yet authorized the detonation, which would take place about 180 miles from the Soviet Union. The guessing is that if it occurs at all it will be delayed for at least two more years. The Geneva nuclear conference and the international picture generally are influencing factors."

The ultimate obstacle to the Chariot project came at the Department of the Interior, the official protector and benefactor of American native populations (although it has often bungled the role). Early on, Interior's Bureau of Land Management had approved withdrawal of the federal land for the Chariot project from public use and access, putting it under AEC control. In July 1961, the Point Hope Eskimo village wrote to Kennedy administration Interior secretary Stewart Udall, protesting the land withdrawal. The village argued that the 1958 federal act granting statehood to Alaska gave the natives "proprietary rights" to federal land superior to the authority of the BLM. At the same time, the Alaskan opponents to the project were keeping Interior informed of what they viewed as the AEC's hostile and high-handed actions and attitudes in the North.

In August 1961, Udall indicated, without a formal finding, that the natives had rights to the land that the AEC must respect. In January 1962, Udall told the AEC that Interior would review any environmental analyses of the Chariot project. That effectively signaled the end of the project, as it became evident that neither Udall nor the White House was likely to let the project move forward. In April 1962, an article in *Harper's Magazine*, "The Disturbing Story of Project Chariot," by Paul Brooks and Joe Foote, brother of Alaska environmental scientist Don Foote, drew further national attention to the stalled Chariot.

The August 24, 1962, edition of the Fairbanks *Daily News-Miner* carried the headline "PROJECT CHARIOT CALLED OFF." The paper printed an Associated Press dispatch, datelined Washington, written by veteran AP reporter Raymond Crowley. His conclusion was both condescending and racist, and no doubt reflected the views of many in the Plowshare program and at the AEC: "Alaskan Eskimos won a victory over atomic science today. The great white father isn't going to order any time soon, if ever, a big nuclear boom on their happy hunting grounds. The Atomic Energy Commission has shelved long-laid plans to blast out a new harbor above the Arctic Circle, near Cape Thompson in northwest Alaska. These plans—known as Project Chariot—had disturbed the Eskimos no end."

13. Sedan Side Trip to Nevada

By the end of 1961, the AEC was internally acknowledging that the Chariot program's environmental studies were damaging the agency's credibility. In May 1962, Livermore told the AEC it was time to park Chariot, which the *New York Times* dutifully reported as a likely outcome, although the lab remained fully committed to the parent Plowshare program.

John Kelly, the AEC's overseer of Plowshare (although Teller didn't take to oversight well), was already planning to limit the publicity damage from canceling Chariot in the face of local opposition and Interior Department obstruction. Kelly concocted an "Information Plan" for political damage control. He suggested that releasing the information that the AEC was canceling Chariot should be "carefully timed to coincide with a convincing event" that would demonstrate that Chariot was no longer technically necessary. Another key part of the damage control would substitute underground explosions for the above-ground blasts that would have been central to Chariot's harbor-blasting.

That carefully-timed event was called Sedan, designed as a huge shot at the Nevada Test Site some ninety miles south of Las Vegas set for July 6, 1962. In order to gather engineering and physics data for the planned Panama Canal project, Sedan was designed as an underground explosion that would create an above-ground crater. It was, therefore, neither atmospheric fish nor underground fowl in the terms that defined the limits of test ban treaties.

The AEC planned the event for its public relations value as much as for what the engineers, geologists, and physicists would learn. Sedan would create a large, visible crater, providing much of the data the scientists hoped to gather in Alaska. Instead of a series of explosions, as planned for Alaska, Sedan would use just one bomb. The shot would also take place at the test site, where there was a substantial buffer between the site and privately-owned land. The test site also had been hosting nuclear blasts for a decade, surrounded by rural ranching and farming communities that generally accepted the AEC as a worthwhile neighbor.

The AEC had to move fast to distract attention from the wheels coming off of the Chariot project. Livermore suggested scrubbing Chariot on April 30, 1962. The White House National Security Agency approved Sedan as a replacement on May 8, with the shot scheduled for July at Area 10 in Nevada. This quick development (warp speed for government work) suggested that the AEC and Livermore had been planning the Chariot successor for some time prior to the official paperwork.

As a first step in the AEC's public relations plan—based on the assumption that everything would go swimmingly—immediately after the explosion, the commission would announce that Sedan had been a great success. The press release, of course, was written well in advance of the event it described. Because the AEC controlled all access to the test site and all information coming from it, the commission could claim success without contradiction. Then the AEC would follow with a press release saying that the data from Sedan "largely obviated the need for Project Chariot."

Physicist John S. Foster, the fourth Livermore director (following York, Teller, and Brown), wrote in a draft letter formally killing the Chariot project: "Perhaps never stated, but of course always kept in mind, is another purpose of Chariot, namely…To dramatically demonstrate (a) the capability of nuclear explosives to excavate large volumes of earth in a controlled and constructive manner, and (b) that such excavations can be done safely with minimal hazard." Those unstated purposes carried over to the AEC's wishes for the Sedan shot. The wishes became the doctrine.

The lynchpin in Teller's entire Plowshare enterprise was a so-called clean bomb, one that released little radiation and lots of explosion. Fission bombs produced a nasty alphabet soup of radioactivity. The splitting of uranium atoms yielded not only atomic fragments, but many additional elements, such as radioactive plutonium, cesium, strontium, and iodine, all of which endanger human health in various degrees and over various time periods. Radioactive strontium, for example, poses little immediate danger, but is long-lived and concentrates in bone. Iodine represents an immediate threat, particularly to the developing thyroid glands of children, but loses its radioactivity fairly quickly, over a matter of days.

Fusion bombs, on the other hand, didn't produce radioactive elements. They didn't split heavy atoms but released energy by combining light (hydrogen) elements. But fusion bombs needed fission bombs as triggers. Teller's quest was to minimize the radioactive effects of the fission trigger as it set off the "clean" fusion explosion. He never succeeded.

On July 6, 1962, Livermore scientists fired off their thermonuclear bomb buried 635 feet under the soft alluvial desert at Area 10. The device had the explosive force of 104 kilotons, equivalent to one hundred thousand tons of TNT. The blast was the largest ever in North America at the time, making the bombs that fell on Japan in 1945 look puny by comparison.

There was a purpose to the large size of the Sedan project. Livermore's Gary H. Higgins, a division leader at the lab, noted in a January 1970 letter to Robert E. Miller, manager of the AEC's Nevada Operation Office, "Comparison of projected costs of nuclear excavation and conventional excavation show that nuclear excavation is only practical and economic when the individual explosive yields are in the order of 100 kilotons and larger." Higgins added, "If the AEC or the U.S. government cannot, in good faith, assert that a useful excavation will result from a project, very few (if any) users will consider nuclear excavation as an alternative to more costly but proven conventional techniques."

The Sedan shot went off at 10:00 a.m. One description said the blast triggered "a rumbling base surge that extended outward from ground zero in a wavelike motion in the earth's surface." A spherical dome six hundred feet across rose from the ground. As it reached a height of three hundred feet, the dome blew apart, hurling rocks and dirt nearly half a mile into the air.

Unlike the early public relations plans for the blast, there was no press presence when the engineers ignited Sedan. Nor were there any outside photographs, although the agency took its own still and motion pictures, as it did of all its blasts.[62] Nonetheless, Sedan got considerable press cover-

---

62 There is a short, impressive AEC film of the Sedan event on YouTube (http://www. youtube.com/watch?v=T1o38Yo5OhY) and is part of a longer documentary available on the web sit for this book, www-toodumb.org.

age, relying on the AEC press releases for information. The United Press International wire service reported, "The mightiest nuclear blast within the United States and the first known detonation of any hydrogen type of explosive in this country was set off underground today. It tore a great open-faced crater in the desert floor."

That massive crater today is visible at the test site. It is a 320-foot deep hole, some 1,200 feet wide, about what the Livermore engineers had predicted prior to the detonation. The next day, the AEC, in a press release, said the blast had ejected some 7.5 million cubic yards of rock and dirt, a "significant contribution to nuclear earth-moving technology." That scientific claim, it turned out, was false. As Livermore's Gary Higgins wrote in 1970, "Sedan, in desert alluvium, does not give information relevant to most of the proposed excavation applications." But the AEC later used the claimed results of the Sedan test with great success.

Within an hour of the Sedan explosion, the AEC issued a press release claiming that "95 percent of the radioactivity was trapped in the ground, or in the earth that fell back promptly." The remaining radioactivity, said the AEC, "had fallen close to the test site." There was, of course, no way for AEC scientists to know those facts so soon after the event.

The alleged success of Sedan turned out to be a tissue of AEC lies; the reality was much different than the press release hyperbole. In 1965, the AEC's final report on Sedan acknowledged that the blast had been much dirtier than expected. The 1965 analysis said, "The Sedan detonation gave a dust cloud which was 50 percent larger than predicted and deposited nearly five times more radioactivity than predicted. This indicates that the relationship between depth of burst, yield, and radioactivity escape used to extrapolate to the Sedan event must be re-examined."

The radioactive plume rose two miles into the air, quickly getting into the upper atmosphere where it could be easily transported great distances. The radioactive cloud began moving north and northeast at 12 mph. Even as the AEC claimed that the test had been clean, events on the ground demonstrated that the agency was prevaricating. The site of the Sedan test, ground zero, was only ten miles from the test site's northern border, so the fallout from the blast was surely already off the site and into civilian territory by the time the AEC put out its glowing pre-written press release.

At 2:45 p.m. the same day, the AEC told reporters that the radioactive cloud had crossed Nevada State Highway 25 some thirty miles from the test site, and state police had closed the road and evacuated people from nearby ranches. The radioactive dust cloud forced the city of Ely, Nevada, well over a hundred miles away, to turn on the street lights at 4:00 p.m. Within five days, the cloud had passed over six states and entered Canada.

On Saturday, July 7, Robert Pendleton of the University of Utah's radiological health department and twenty students were reading background radiation some twenty miles southeast of Salt Lake City, looking at the presence of radioactive cesium. A red-brown cloud approached the crew from the direction of the Nevada Test Site and passed over Pendleton and his students. "Our instruments went completely nuts and we couldn't measure what we were trying to do," he reported. Pendleton and others from the university began collecting samples the next week of milk and fodder from farms near Salt Lake, finding high levels of radioactive iodine, which they attributed to Sedan and other tests at the test site. Pendleton advocated keeping the milk out of the liquid market, diverting it to production of cheese and powdered milk, giving time for radioactive iodine to decay. The AEC and the U.S. Public Health Service ignored his data and warnings. Later, AEC scientists concluded that they had badly underestimated the threat of short-lived radioactive elements such as iodine-131, with an eight-day half-life, contained in radioactive fallout. Soon, scientific disputes over radioactive fallout would explode on the scene.

A week after the Sedan explosion, Nevada Highway 25 was still closed and radioactive debris had to be hosed off with pressurized water. For the next eleven years, during which approximately twenty more shots were fired, radiation proved to be a problem neither Livermore nor the AEC could control. Teller biographer Peter Goodchild noted, "The 'clean' bomb, so much the basic essential of the Plowshare vision, was to remain an enticing chimera forever on the horizon, one of Teller's motivations in his continuing fight against the test bans. To this day, it has never been fully realized."

The Sedan explosion left one tangible, lasting result—a hole in the desert. The bomb blew 12 million tons of Nevada Test Site earth into the air, leaving the largest crater ever made by man—a giant dimple in the earth some 390 yards in diameter and 97 yards deep.[63]

The reality of the Plowshare failures never dented Teller's atomic optimism or that of the AEC program overseers. Teller remained convinced that it was possible and practical to use nuclear weapons as substitutes for bulldozers and manpower. His personal brilliance and persuasive skills, combined with the AEC's prodigious public relations machine, concealed and glossed over the problems with Sedan and kept funding flowing to the Plowshare program. Geographer and historian Scott Kirsch noted:

> Public representations of the experiment remained highly selective. The degree to which Plowshare advocates were able to "carefully phrase and limit the subject" (as an earlier AEC public information directive had put it) was critical not just to public knowledge of Sedan and the Plowshare program but also to the practice of nuclear excavation as an experimental science carried out within the classified spaces of the Nevada Test Site and the Lawrence Radiation Laboratory at Livermore. For despite Plowshare's apparent defeat in Alaska, the emergence of disturbing new arguments about the health hazards of ingested radioiodine, and ostensibly severe limitations imposed by the Limited Test Ban Treaty of 1963, its excavation program actually gained unprecedented funding increases for nuclear cratering tests between 1962 and 1968.

A positively glowing article in *Life* magazine in March 1963 illustrates the force of the atomic publicity apparatus. The article began with a lovely underground color photo of a cavern created by the 1961 Gnome test in Carlsbad, New Mexico, apparently taken by an AEC photographer—since there was no source given for the photo (or for the black-and-white photos of the Sedan shot that accompanied the article's text). The *Life* article said, "Presenting these Project Sedan results to Congress in its annual report, the AEC predicted that it would probably be able to take on major

---

63 Tourists can visit the Sedan crater, south of Las Vegas, on scheduled trips leaving from the gambling mecca.

earth-moving projects in about five years." The article added an entirely misleading veneer to its report, no doubt provided by the AEC. "In the underground Sedan blast, more than 70 percent of the energy released was 'clean,' i.e., it did not contaminate the surrounding earth with fission products."

A detached observer might have seen the Sedan explosion as a warning to the AEC and Livermore scientists and engineers that H-bomb excavation might be a futile and dangerous venture. But optimism blinded even the smartest Plowshare advocates. AEC chairman and Nobel laureate Glenn Seaborg came close to exposing Sedan for what it was in an undated presentation—most likely given to the congressional Joint Committee on Atomic Energy in 1970—included in a published 1970 compilation of his testimony and speeches. Seaborg said, "Our experience with nuclear cratering has been limited. A large number of physical processes take place concurrently and successively in a nuclear excavation explosion, and these processes proceed almost instantaneously. Therefore measurements are very difficult; there is little wonder that nuclear cratering has not yet advanced to the state of a precise technology."

How about those polar-bear shaped harbors, Dr. Teller?

14. A Man, a Plan, a Canal

Having struck out in Alaska and hitting a foul ball on the Nevada Test Site, the AEC turned its attention southward, toward an objective Teller and others had been thinking about for years. As the palindrome has it: A man, a plan, a canal, Panama. Add a series of H-bomb explosions, and Teller's vision came into focus: a sea level passage between the Atlantic and Pacific oceans, rendering the existing canal, heroically built between 1904 and 1914, redundant. Ultimately, like all the other ambitious Plowshare projects, the Panama Canal program failed—a victim of technical hubris, practical obstacles, and daunting international problems, including fundamental instability in the government of Panama and problems erected by the Limited Test Ban Treaty of 1963.

The Panama Canal—one of the engineering wonders of the world— opened in August 1914, just as World War I was getting underway in Europe. The ship channel across the Isthmus of Panama changed patterns of commerce and military response worldwide, dramatically increasing international business while reducing risk and dramatically expanding the scope of U.S. military power. The canal, which connected the Pacific Ocean[64] and the Atlantic, cut weeks and millions of dollars off the costs of shipping goods between the two oceans. It also meant that shipping didn't have to run the nasty, twisting path at the bottom of South America, rife with lousy weather, narrow channels, icebergs, and long distances down and up the massive continent.

Historian David McCullough, in the preface to his fine work *The Path Between the Seas*, wrote: "The creation of the Panama Canal was far more

---

64 Because of the geography, the Pacific at the canal site is actually east of the Caribbean outlet to the north.

than a vast, unprecedented feat of engineering. It was a profoundly important historic event and a sweeping human drama not unlike that of war. Apart from wars, it represented the largest, most costly single effort ever before mounted anywhere on earth."

The U.S. Army Corps built the Panama Canal at the beginning of the twentieth century and retained an interest in it, including stationing troops in Panama, in the U.S.-owned Canal Zone until Panama took full control in the end of the century. The Canal Zone was actually U.S. territory, subject to U.S., and not Panamanian, laws.[65] Until November 1903, the isthmus was part of Colombia, but a local revolution, with the backing of the United States, led to creation of the Republic of Panama and the signing of the Hay-Bunau-Varilla Treaty. The treaty granted the United States a ten-mile-wide swath of territory from Panama City to Colon for the construction and operation of the canal. A U.S. firm, the Panama Canal Co., took over French interests to run the canal and collect the considerable tolls.

President Theodore Roosevelt's secretary of state, John Hay, who negotiated the treaty with the new Panamanian republic, wrote at the time to Wisconsin Republican Sen. John Coit Spooner, that the treaty was "very satisfactory, vastly advantageous to the United States, and we must confess, with what face we can muster, not so advantageous to Panama....You and I know too well how many points there are in this treaty to which a Panamanian patriot could object." The awareness of the unbalanced nature of the treaty would later have an impact on the plans for blasting out a new canal.

Ships enter the canal from the Atlantic north of Colon and travel 51 miles, typically a nine hour journey, to the Pacific Ocean entrance at Balboa. But the canal is not perfect. A major problem is getting over the continental divide, which separates the Atlantic and Pacific watersheds. That requires locks to lift vessels some 85 feet. The canal has three sets of locks, with two locks at each set. That causes backups at the entrances

---

65 Arizona Republican Sen. John McCain, the 2008 GOP presidential candidate, was born in the Canal Zone. Had Panama owned the canal at the time, McCain may have been ineligible to be president, as the U.S. Constitution arguably requires birth in the United States for presidents.

to the locks on the canal. In shipping, as in much of business, time is money.

The size of the locks also limits the size of the ships, both commercial and military, that can pass through the engineering marvel. The Bridge of the Americas at the port of Balboa, where the canal links to the Pacific, also limits the height of the vessels that can pass to under about 190 feet. The width of the locks is about 110 feet, with a useful length of 1,000 feet, defining the parameters for what are known as Panamax ships. As a practical matter, these are ships that can carry about 70,000 tons of cargo.

The United States had long recognized the value of a sea level canal bridging the Caribbean and the Pacific. The initial plan for the Panama Canal was for a sea level project. That proved impractical, and the idea was abandoned in 1906 in favor of a lock-based project. But the dream of a sealevel isthmian canal never completely faded. In 1947, Congress approved a series of studies of thirty sea level canal routes covering a wide swath of territory from Ecuador to Mexico. Among other drivers for a canal upgrade or new canal was military. While many analysts had talked about a third set of locks to overcome the size limits of the canal, military experts believed that locks were particularly vulnerable to attack from air and sea. A second canal, at sea level, would be less vulnerable and would allow military shipping to pass from the two oceans faster.

The early studies found the most promising route was in the Darien region on the Caribbean coast near the border with Colombia. In October 1956, the United States Joint Chiefs of Staff decided that a sea level canal was not possible politically, although militarily desirable. The Plowshare project in Panama would once again raise the vexing question of what is military and what is civilian.

The AEC and the Panama Canal Co., a U.S. firm which operated the canal from its inception until 2000 when the government of Panama took over canal ownership and management, had discussed using nuclear bombs to excavate a new canal as early as 1947, although nothing came of those low-level musings for a decade.

In the February 1957 Livermore conference to launch nuclear excavation and landscaping, one of the most popular ideas that arose was to use a series of nuclear blasts to create a new canal across the Isthmus of Panama.

The secret meeting included twenty-four scientists from Los Alamos, Livermore, and the rest of the vast U.S. atomic energy complex.

A paper at the 1957 conference predicted that the existing canal would face devastating slowdowns after 1961. A *Wall Street Journal* article in March 1961 noted a "canal bottleneck" and predicted that the result could be to "throttle the expanding international trade of the United States and a score of other nations." The Livermore paper estimated that building the canal across the preferred route would require exploding twenty-six nuclear bombs, totaling 16.7 megatons (equivalent to 16.7 million tons of TNT). To deal with the radiation problem, the builders would have to relocate the local population, mostly indigenous Indians, for an unspecified time.

The idea for a canal created by nuclear explosions began not in Panama but in Egypt. The Suez Canal, connecting the Mediterranean with the Red Sea and opening a short sea route between Europe and Asia, opened in 1869. By the mid-1950s, the canal had become a pawn in international politics. Egyptian dictator Gamal Abdel Nasser was building the Aswan High Dam to contain the Nile for flood control and electric generation. Nasser had financing from the United States and Great Britain for the dam, but both rich countries withdrew their support when Nasser, playing Mideast power politics, started cozying up to the Soviet Union.

In 1956, Egypt nationalized the canal, which had been run by the Suez Canal Company, a private Egyptian company with heavy French investment. Nasser hoped to use the revenue from the canal to cover the loss of the French and British investments in the Aswan dam. In response, France, Britain, and Israel invaded Egypt. Using mines and sunken ships, Nasser blockaded the canal. With U.S. backing, Canadian prime minister Lester Pearson intervened, persuading the United Nations to create a multinational peace-keeping force, which restored stability to the canal and the region.[66]

In the United States, Edward Teller saw the events in Suez and envisioned a way to resolve the crisis permanently: a second canal located in Israel. In a 1968 textbook, Teller writes that during the Suez crisis in mid-1956, he and three colleagues met at Livermore, where they "raised the possibility of cutting another canal through friendly territory with

---

66 Pearson won a Nobel Peace Prize for his efforts.

nuclear explosives." For political and practical reasons, nothing came of this idle Suez speculation, but it planted a seed in the minds of Teller and Livermore's Gerald Johnson, later an important Plowshare official.

The Panama project fired imaginations at the 1957 Livermore conference that created the Plowshare program, but it had to take a place behind the program's more mundane plans for domestic blasts, particularly Chariot and later Sedan. One of the rationales for Sedan was to generate data about the way nuclear explosives would create craters that could be connected to form a long trench as the basis of the new canal. While the Livermore engineers and scientists learned some valuable facts about cratering from the Sedan explosion, there was much more data needed before a practical approach to a Panamanian canal could begin.

As Livermore and the AEC began studying the basics of building a new canal, the Panama Canal Co. hired the engineering firm of Parsons, Brinckerhoff to study the feasibility of a third set of locks or a new sea level canal. According to Panama Canal historian John Lindsay-Poland, officials of the San Francisco–based engineering company read about the AEC's Plowshare program and the plans for Alaska and contacted the San Francisco Operations Office about their studies for the canal company. In July 1958, officials from Parsons met with AEC commissioner Willard Libby, who assigned Livermore to analyze the nuclear option for the study, completed in February 1959. Libby told the Parsons engineers that the blasts would require evacuating some populations and establishing a twenty-five-mile buffer zone between the project and any significant populations, effectively ruling out the existing site of the canal.

Livermore's notion to span Panama with nuclear explosives had a solid ally at the AEC's Division of Military Applications in Washington. The division chief was Gen. Alfred Starbird of the Army Corps of Engineers.

Starbird told Gerald Johnson, whom York and Teller had put in charge of the Plowshare program, that he should involve the Corps in the Panama project. Livermore's blasting plan already had the theoretical endorsement of the Panama Canal Co. Johnson later said, "The Panama Canal Company interest and the Corps of Engineers' interest is really what put the heavy focus on excavation. Because they, in a sense, were a potential customer. We had a target and a high level target." Teller and Johnson briefed President Eisenhower and his cabinet on the canal and other Plowshare projects.

Despite the work of the Army Corps in 1947 in scoping out alternative routes, the U.S. government was considering routes outside Panama in 1958. One alternative, which Eisenhower favored, was the Mexican Tehuantepec passage. Ike considered publicly announcing U.S. interest in it, but Christian Herter, his secretary of state, dissuaded him. Herter raised the harmful impact such an announcement would have on Panamanian relations. The Livermore analysis, presented to Johnson in March 1958 (in carefully couched language so as not to displease the White House), found lots of obstacles to the Mexican route for a new canal, concluding: "It should be realized that a great deal of additional planning would be necessary before any actual construction work could be started." Johnson wrote on the cover page, "Good report." That was the last gasp of a Mexican Atlantic-Pacific canal.

In January 1960, the Panama Canal Co. began reviewing and updating its earlier studies done with the AEC on a new canal. In early 1962, President Kennedy asked the AEC and the Army Corps to look at the feasibility of using nuclear bombs to create the new canal, leading to a series of studies in 1964, after Kennedy's assassination. Using H-bombs to build the canal made sense only if it was considerably cheaper than conventional digging and blasting. The 1947 studies, done by the Corps, which identified the Darien route as the most practical, put the cost of a conventionally-built canal at $5.1 billion. It was never clear that a Plowshare project could come in under this figure. As the commission began its work, challenges to the economics of nuclear excavation began to surface. In a 1962 speech to the American Nuclear Society and a 1963 paper based on the ANS presentation, Teller argued that the cost of nuclear explosions in the megaton range would be two cents per cubic yard of earth moved, versus one dollar for conventional excavation. That figure drew no challenges, and early studies seemed to accept at face value AEC claims of the economic cost savings from nuclear explosions. A 1960 Army Corps study conducted with Livermore concluded that the Darien route would cost $770 million using nuclear excavation technologies, versus the 1947 estimate of $5.1 billion using conventional excavation methods.

But Army Corps Col. James Stratton, who worked on the Corps' 1947 studies, raised caution flags about "dubious" cost estimates. In an article in *Foreign Affairs*, Stratton noted that AEC chairman Glenn Seaborg had told

Congress that verifying the efficacy of nuclear excavation technology would cost some $250 million. Stratton argued that at least part of those research and development expenses should be charged against the sea level canal.

Stratton also warned that the existing estimates did not incorporate the costs of infrastructure, including protecting the human population along the route, or the special costs of military defense. A full-in estimate, he suggested, would bring the cost of the nuclear canal up to about the equivalent of conventional technology. If that was the case, he said, using nukes to blast out the canal route was foolish.

A major international obstacle also faced plans for a nuclear blast across Panama, a hurdle that bedeviled Teller throughout Plowshare's lifetime. In October 1963, the United States ratified the Limited Test Ban Treaty (much to Teller's chagrin). The treaty prohibited nuclear explosions in the atmosphere, the oceans, and space. The Panama project would feature underground blasts, which were not out of bounds under the treaty. However, the treaty also banned any underground shot that would have caused "radioactive debris to be present outside the territorial limits of the state under whose jurisdiction or control the explosion is conducted." Given the small size of Panama, the geography of the isthmus, and the best route for a sea level canal, snuggled up next to Colombia, it would have been impossible to conduct a series of blasts the size of those contemplated for the canal without dropping fallout on Colombia and half a dozen or so other Central and South American countries.

The next blow to the canal project came when President Johnson signed a law in September 1964 creating the Atlantic-Pacific Interoceanic Canal Study Commission to mount an investigation "to determine a site for the construction of a sea level canal" connecting the oceans. Congress appropriated $17.5 million for the exploration of the sea level canal. That appeared to be a positive development; it was not to be.

As the Army Corps began field studies in Panama, the project ran into a series of technical, economic, and political complications, as well as difficult interactions within the convoluted web of U.S. federal government agencies and contractors. One assessment said that the studies involved thirteen federal agencies, including six military agencies, twenty-seven prime contractors, and up to one hundred seventy Canal Commission personnel located in Darien. Much of this activity was designed to mollify the government of

Panama, which was facing pressure from leftists in the country to take control of the canal away from the Americans. Following riots in Panama and the Canal Zone in January 1964, which saw twenty-four Panamanians and four U.S. soldiers killed, Panama broke off diplomatic relations with the United States. Panamanian president Roberto Chiari demanded renegotiation of the 1903 canal treaty. The countries reestablished diplomatic relations in April 1964. The study commission was designed in part to paper over the tensions between Panama and the United States. In December President Johnson agreed to negotiations on a new treaty.

While the White House may have seen the nuclear canal project in the context of international relations, in Congress, and particularly in the all-powerful Joint Committee on Atomic Energy, the canal project had emerged as an important symbol of American muscle. Sen. John O. Pastore, the Rhode Island Democrat who chaired the committee in 1965, said that the canal was "the one thing that has given this thing life and the one thing that has more or less enthused this committee to provide the money for Plowshare. Once you have ruled that out, I am afraid interest is going to drop off."

In addition to the test ban treaty, another international agreement threw sand in the slowly-meshing gears of the nuclear canal. Negotiations started in 1965 for the Treaty of Tlateloco, which was signed in February 1967 and went into effect in April 1968. The treaty banned nuclear weapons in Latin America, although there was an ambiguous provision that the AEC interpreted as permitting the use of H-bombs in civilian construction. At the same time, the existing conundrum of how to distinguish so-called peaceful uses of explosives from weapons continued. As physicist and journalist Barbara Levi wrote in an analysis for Princeton University, "The potential of nuclear explosions for both destructive and beneficial uses has posed a persistent arms-control dilemma: what measures can be taken to deny nuclear weapons to a nation without also denying it the possible benefits of peaceful nuclear explosions?"

As part of the Panama Canal program, Livermore and the AEC were setting off a series of nuclear explosions at the Nevada Test Site to explore various technical issues, including the effect of soil characteristics, ground shock, air blast effects (such as blowing out windows), and the spread of radiation. Two of those tests, known as Sulky and Palanquin, demonstrated just how little the nuclear scientists actually knew about controlling their technology.

15. The End of the Exploding Game

The failures of Palanquin and Sulky marked a quiet end to the Panama fantasy. It died with a whimper, followed by the entire Plowshare program, which fizzled out in the early 1970s with no fanfare. Despite repeated assurances from the White House that the Panama Canal project was on the government's priority list—those statements were convenient fiction that served the needs of the administration to feed the nuclear enthusiasts in the Congress's Joint Committee on Atomic Energy.

President Johnson said on television on December 18, 1964, that the United States was seriously interested in blasting out a canal in Panama using nuclear bombs. The same day, Livermore lit off the Sulky blast, a small[67] explosion in a test site shaft at Area 18, Buckboard Mesa. Livermore designed Sulky to examine the effects of cratering and extent of radiation dispersal in hard, dry rock. They were also designing it as a substitute for the larger Schooner excavation blast, with the size terms of the Limited Test Ban Treaty ratified in 1964 in mind. The boffins expected that burying the Sulky shot deeper than originally planned would produce a larger crater, reducing the radiation that escaped from the explosion. But Sulky didn't create a crater. Rather, it produced a broken rock pile. What did the scientists and engineers learn? As historian and geographer Scott Kirsch put it, the Plowshare program learned that "if nuclear explosives were buried too deeply, these data had shown they made mounds instead of craters."

Palanquin followed Sulky. Palanquin was also relatively small, four kilotons equivalent, also an attempt to solve the vexing fallout problem while skirting the test ban treaty. The idea behind Palanquin, Kirsch explained, was "down-hole deep debris entrapment. If the engineers could direct the

---

67 Equivalent to ninety-two tons of TNT, or 0.92 KT.

radioactive dirt and dust from the explosion downward into a cavity cre
ated by a June 30, 1964, blast, dubbed Dub, they would have created the
equivalent of the long-awaited 'clean bomb.'" The idea was so promising
that Livermore moved up the Palanquin test in the queue of explosions the
Firecracker Boys hoped to set off.

Livermore did not expect much radiation from Palanquin, so the scientists limited monitoring and sampling plans too close to the site. The
AEC also pulled the cone of silence down over Palanquin. An AEC directive ordered: "Every effort will be made to prevent the public from becoming aware that Project Palanquin is scheduled for execution soon." The
agency expressed anxiety about the perception that the test might violate
the treaty. The directive continued, "In particular, we want to avoid misleading or speculative articles which might create the impression that the
United States thought it was violating the terms of the nuclear test ban
treaty. To the contrary, the United States does not think it is or it would not
have authorized conduct of the experiment."

Palanquin went off course from the beginning. Milliseconds after the
detonation on April 14, 1965, radioactive gases spurted high into the air
as the grout failed that sealed off the bore hole where the bomb was buried.
The pressure of the explosion was supposed to push the radioactive gases
down into the Dub cavity. Instead, the gases jetted through the four-foot
hole, along with ten times more water vapor than the scientists expected.

Even dull government language conveyed the wild ride of Palanquin.
A June 3, 1965, physics report on the blast, declassified in 1994, reported:

A "plug" of material was ejected from the emplacement hole, causing a
break in the cavity and consequently a sudden drop in cavity pressure.
This stemming breach created a high velocity jet of hot gases and rocks
which led to an "erosional" crater similar to that produced by gas or
steam-well blowouts. Essentially all of the radioactivities were first in
gaseous form; then most of these activities condensed in fragments of
vesicular glass varying from a millimeter up to several tens of centim-
eters in size. On the average, only about 1 percent of the radioactive
fragments passed beyond a range of five miles, although some radionu-
clides of a volatile nature were much more abundant at longer ranges.

The eye-witness report continued: "During the first ninety milliseconds, cavity pressure was high enough to rupture the thirty-six-inch emplacement casing (three-quarter-inch wall thickness, and a free air burst strength of eighty to one hundred bars)."

The result was a plume of radioactivity that reprised the catastrophic Sedan fallout track. Surprised government officials, including AEC chairman Glenn Seaborg, began tracking the radioactive path of the Palanquin explosion, but issued no public statements. As was typically the case, the AEC decided on silence in order, it said, not to cause fear in the public. That decision also reflected a desire to not inflame public opinion against future testing. Seaborg, who was keeping the White House updated, was fearful that the Palanquin cloud would pass over the U.S. border and into Canada, violating the Test Ban Treaty.

Within a day, the Palanquin plume had passed east of Spokane, Washington, and was headed toward Canada. Soon, it was over Butte, Montana, and excess radioactivity had passed into Canada. Seaborg said the radiation levels were so low it was "doubtful whether anyone else would detect them." Additionally, in an excuse any parent would recognize, when the agency publicly announced the test four days later, the AEC and the White House whined that they had not violated the treaty and, even if they did, it wasn't as bad as what the Russians did in January in a test that spread radiation to Japan.

While the radiation diminished with distance and dilution, local radiation near the test site was far higher than the scientists had predicted. According to AEC records, six local families were moved to the NTS base camp for whole-body radiation counts and four of the families, including eighteen children, showed "slight thyroid burdens" from radioactive iodine concentrations. AEC officials later acknowledged that predicted fallout levels from Palanquin, which was designed to contain the fallout underground, were "low by a factor of one hundred at five miles, low by a factor of thirty-three at ten miles, low by a factor of ten at eighteen miles, and low by approximately a factor of five at forty and fifty-five miles."

Palanquin marked the beginning of the end for Plowshare's excavation and cratering program. The firecracker office had three more explosions planned in the cratering test series: Cabriolet, a small (2.3-kiloton) blast at the Nevada Test Site; Buggy (a row of five 1.1-kiloton bombs) on the test site; and Schooner, a much larger (35-kiloton) blast. The Plowshare program had originally designed Schooner as a 100-kiloton test on federal land at the eastern Idaho reactor test site, scheduled for 1966, but was unable to bring it off following Palanquin's problems. A month after the Palanquin fiasco, Livermore's Glenn Werth vowed that Schooner would continue as Plowshare's top cratering priority.

But Palanquin's failure put all future Plowshare blasts on hold, on orders from Lyndon Johnson's White House. The *New York Times* quoted Deputy Defense Secretary Cyrus Vance[68], who said the White House did not want the tests to open "another propaganda front" in the public relations battle with the Soviet Union. Reporting at a November 1968 Plowshare meeting in Las Vegas, the AEC's H.D. Bruner noted, "The Palanquin failure set back the program by two years." The White House cleared the AEC to go forward with Cabriolet and Schooner. On January 28, 1968, the Cabriolet explosion went off, an event one historian described as "another low-yield, low-publicity test in Nevada." Buggy followed on March 12, producing a trench nine hundred feet long, the first, and ultimately only, attempt at what would be needed to dig a canal. The always enthusiastic *New York Times* headlined its story, "Buggy, Cabriolet A-Test Prove Canal Feasibility." That was, it turned out, the usual overstatement.

The long-delayed, considerably-downsized Schooner test followed on December 8, 1968, not in Idaho as planned, because of local objections, but on the test site. The 35-kiloton bomb, buried three hundred fifty feet deep in hard rock in the test site's Area 20, blasted a crater eight hundred fifty feet across and two hundred feet deep. The AEC, in a public statement two days later, claimed that the radioactivity from Schooner was "infinitesimal," not quite a hard number. Robert Pendleton, the University of Utah scientist who with his students had been showered by Sedan and first raised the alarm about iodine-131, claimed the Schooner fallout would increase risks for cancer from long-lived radioisotopes.

Pendleton acknowledged that Schooner did not produce the nasty radioactive brew that rained down from Sedan. The AEC's claims for Schooner, Pendleton said, were "gobbledygook" and that a "credibility gap between

---

68 Later Jimmy Carter's secretary of state.

the AEC and the knowledgeable people of the western United States" had opened up. Indeed, the AEC did its best to stifle information about off-site radioactivity from Plowshare tests. Chairman Seaborg rejected a draft of the AEC's annual report to Congress, complaining of "too much detail on radiation off-site."

Seaborg also made sure the AEC report did not mention that fallout from Schooner produced an egregious violation of the test ban treaty, although the Canadian Broadcast Corporation reported that sampling at some monitoring stations December 13 to 15 were ten to twenty times normal. For reasons of cold war solidarity, the Canadian government made no protests over Schooner's border-crossing atomic trespass.

One powerful local had begun to doubt the AEC's radiation estimates. The reclusive, obsessive billionaire Howard Hughes, who lived in Las Vegas and offered to fund further research into what was going on at the test site. A *New York Times Magazine* article by science reporter Gladwin Hill blamed AEC secrecy for the credibility gap. "The AEC originally tried to conduct the testing here in complete secrecy," Hill wrote. "The result was the nationwide furor in the 1950s over fallout hazards. The AEC was forced to shift to a policy of candor, which endured until testing went underground. At that point, the man who is still the AEC's chairman, Dr. Glenn Seaborg, dually distinguished for his knowledge of physics and his lack of knowledge about public information, imposed a virtual news blackout on testing details except for terse official communiques." Hill added, "Since Howard Hughes spoke out, the AEC has been scrambling to get back on the candor wagon and spread the word of its safety policies and practices, obscured in the news blackout."

The radioactive curtain started to discernibly descend on the Plowshare program's Panama pretentions with the failure of the Carryall project. Carryall, on the Plowshare plate since 1963, was a mammoth, 1.8-megaton plan to blast a pass through the Bristol Mountains in California's Mojave Desert. It would create an easier East-West passage for the Santa Fe railroad and for the nation's legendary cross-country highway, Route 66. The project was one of Edward Teller's favorites. Developed by the California Division of Highways (under Gov. Pat Brown[69]), the railroad, and Livermore, the project would chop fifteen

---

69 Father of two-time anti-nuclear California governor Jerry Brown.

miles off the railroad route, cutting nearly an hour off travel time and saving the railroad considerable money. Without the nuclear blasts to clear out the mountains, the new route would require a two-mile rail tunnel, rendering the plan uneconomical. The Firecracker Boys would supplant the tunnel by smashing through the two miles with twenty-three bombs, set off in two stages.

Carryall got caught up in the repeated Plowshare delays, with the added weight of doubts about the AEC's estimates of radiation from the blasts. The project feasibility study predicted that Carryall would release six hundred sixty tons of fission products. That low-ball guesstimate was only 66 percent of the fallout from the Sedan blast although Carryall was eighteen times larger. The estimate drew a skeptical review from M.L. Merritt of Sandia Corp., a New Mexico AEC contractor that performed conventional explosions as part of the Plowshare program. Merritt concluded that the Carryall analysis "includes nearly every safety factor that should be considered except the offsite effects of the dust cloud." Merritt also expressed doubts about whether Carryall could comply with the test ban limits.

After multiple postponements, Carryall got a final schedule for the fall of 1968. But the project, far closer to population centers than any of the blasts at the test site, was raising serious objections from an increasingly well-informed public. Focusing on the AEC's habitual secrecy about radiation, noted health physicist C.D. Casoyas of the University of California's San Francisco Medical Center wrote the AEC in May 1967. He charged Livermore and AEC Chairman Seaborg with a plan "to suppress an open discussion of the failure of the Lawrence Radiation Laboratory to assess the number of casualties that the population of Los Angeles would suffer from the low-level radiation effects of the 1.7-megaton Carry-All nuclear cratering blast..." Noting an unexplained delay in Carryall announced in early 1966, Casoyas added, "Up to the present time, neither the Lawrence Radiation Laboratory, nor the State of California, nor the Atomic Energy Commission will state the reasons which led to the postponement of the Carry-All blast, even though its feasibility had been established in 1963, and it would have served the purposes of the Santa Fe Railway System." While the AEC made no formal announcements, Carryall simply vanished from the schedule of Plowshare experiments.[70]

The Panama curtain fell, signaling the end of the show, when the canal commission appointed by Lyndon Johnson in 1964 made its official report to President Nixon in 1970. The report was equivocal, but it didn't take Livermore and the AEC long to read between the carefully-written lines. The ultimate judgment of the canal commission was that a sea level canal didn't make economic, geopolitical, or military sense. As for using H-bombs to blast a sea level canal, the commission report said that "although we are confident that someday nuclear explosions will be used in a wide variety of massive earthmoving projects, no current decision on U.S. canal policy should be made in expectation that nuclear excavation technology will be available for canal construction."

Ultimately, the failure of the Panama Canal project was the end of the Plowshare program. Livermore continued with some plans for the program for another three years, but without any political support in the Nixon administration or the Atomic Energy Commission. Ultimately, faced with across-the-board failure, Plowshare died—not with a bang, but a whimper.

---

70 The project does not appear anywhere in the text of the 2000 official history of Plowshare written by the Department of Energy's Nevada Operations Office, showing up only in an appendix titled "Proposed projects (little work or field testing conducted)," with a project date of November 1963. The closest allusion to Carryall is a comment, "Some of the projects that were never executed had the potential for unacceptable levels of radioactive fallout." The 2000 history also includes what appears to be a deliberate misstatement: "The planned and actual yields of all the Plowshare tests, except for some of the device development tests, were publicly announced in advance of the tests." They were not.

False Scarcity and Fools for Fuels

British grammar school teacher and science fiction writer Henry George Wells in 1913 wrote one of the most astonishingly prescient passages in literary history. In his largely unreadable novel, *The World Set Free*, Wells invents a physicist lecturing to a class in Edinburgh. The physicist, named Rufus, proclaims:

> A little while ago we thought of the atoms as we thought of bricks, as solid building material, as substantial matter, as unit masses of lifeless stuff, and behold! These bricks are boxes, treasure boxes, boxes full of the intensest force. This little bottle contains about a pint of uranium oxide; that is to say about fourteen ounces of the element uranium. It is worth a pound. And in this bottle, ladies and gentlemen, in the atoms in this bottle there slumbers at least as much energy as we could get by burning a hundred and sixty tons of coal. If at a word, in one instant, I could suddenly release that energy here and now it would blow us and everything about us to fragments; if I could turn it into the machinery that lights this city, it could keep Edinburgh brightly lit for a week. But at present no man has an inkling of how this little lump of stuff can be made to hasten the release of its store.

As the United States pursued the panoply of nuclear pipedreams in the 1950s, 1960s, and 1970s, the nation faced what it believed, erroneously, was a limited supply of fuel for the energy revolution that Wells described. In chasing supposedly scarce uranium in order to build its ever burgeoning backlog of bombs, fuel bombers, propel the nation into space, and rearrange the world's geography with fission and fusion, the Atomic Energy

Commission and the Joint Committee on Atomic Energy set off a uranium rush in the U.S. Southwest with attendant boom-time effects. The inevitable collapse left miners unemployed, communities abandoned, and large piles of radioactive detritus from the mills that converted ore into yellowcake (uranium oxide ready to turn into fissionable material).

The government's efforts to find domestic supplies of uranium—by establishing an almost perfectly imperfect market—also set the perfect conditions for market chaos. The AEC's uranium frenzy led to a series of irrational fuel supply contracts, a firestorm of contract cancellations and the largest commercial lawsuit in history. Another unanticipated result was an international cartel formed to corner the uranium market and drive up prices. In the end, the fuel fiasco offered a classic display of government wimping out in the face of an international confrontation, led by the political exemplar of wimpishness—the Carter administration.

The search for technologies to supplant uranium also led to feckless government attempts to invent the equivalent of perpetual motion machines—breeder reactors that created more fuel than they used, as the government repeatedly proved its willingness to spend a dollar today to save fifty cents tomorrow.

16. Uranium Rush and the New '49ers

When the Royalton, Michigan, shop where Vernon J. Pick rewound electric motors burned down in early 1951, the successful small businessman was left with only $13,000 in an insurance settlement. Emotionally as well as physically burned out, Pick, forty-eight, and his wife Ruth packed up a panel truck and a travel trailer and headed toward Mexico for an extended vacation, intending then to migrate to California for a new start in life.

They never made it all the way. When the couple got to Colorado Springs, Colorado, Pick heard about a boom in uranium prospecting that was rolling over the region. So they headed to Grand Junction, west of Colorado Springs on the Utah border, where the Atomic Energy Commission had set up a regional office coordinating the agency's red-hot quest for uranium. Pick made friends with Charles "Al" Rasor, a geologist who headed the AEC's mining office, hoping to pick up some tips about uranium prospecting. Rasor pointed the Michigander to the tiny town of Hanksville in the remote Henry Mountains in the southeastern Utah desert, just north of today's Lake Powell.[71] "If I were going after the stuff," Rasor told Pick, "I'd look here."

That's where Pick looked…and looked…and looked. After nine months of wandering in the desert, mostly in harsh territory where he could only hike in by foot, Pick was ready to give up. He was footsore and sick from drinking the arsenic-laced, alkali water, subsisting on oatmeal and dried milk. Resting on a boulder along Muddy Creek on June 2, 1952, facing a four-day trek back to civilization, Pick notices that his scintillation monitor didn't seem to be working. It was stuck on the high end. He figured

---

71 The Henry Mountains were the last mountain range to be added to the map of the lower 48 forty-eight U.S. states, in 1872.

it had a dead battery. But when he moved away from his perch, the device for registering radioactive disintegrations acted normally. When he moved back to the boulder where he had been resting, the instrument pegged out again.

Pick suspected, hoped against hope, that he might have found uranium ore. When he chipped pieces off the boulder with his prospector's hammer, he found the characteristic tint that indicates uranium ore. "It was all beautiful yellow-orange-colored ore," he said later.

Pick chipped off some more pieces and put them in his pack, stumbled and limped slowly back to his truck over the next four days and soon had the material assayed in Grand Junction. It was high-grade uranium ore. Pick staked his claim, raised money based on the guaranteed high price that AEC was paying for uranium, and was soon mining ore by primitive methods, grossing $50,000 a month. He had struck it rich.

In August 1954, Pick sold his mine to New York–based Atlas Corp. for $9 million in cash and stock, after already taking about $1 million worth of ore out of the mine. At the time of the sale, Pick told the *New York Times* that he remained "interested in the broader prospect of developing uranium as a new national resource for the production of nuclear power and energy for peacetime uses."

Vernon Pick was a beneficiary of an artificial market the U.S. government created for an element that wasn't identified until the late eighteenth century and had little commercial value until scientists realized it could be turned into the most fearsome weapons in history. In its lust to control the supply of uranium in the years during and after the war, the Atomic Energy Commission purposefully built a mining and milling boom that turned portions of the western U.S. desert into the modern equivalent of the California gold fields of the 1850s, creating instant millionaires like Vernon Pick.

From 1789, when German pharmacist Martin Heinrich Klaproth discovered the element uranium,[72] until 1939, when Italian physicist Enrico Fermi proved that the strange metallic element could sustain a controlled fission reaction that released enormous amounts of energy, uranium was a

---

72 Klaproth, 1743-1817, an inventive chemist and pharmacist, also co-discovered titanium and zirconium.

scientific curiosity. When Fermi's pile of graphite and uranium on a squash court under the University of Chicago's Amos Alonzo Stagg Football Stadium demonstrated the controlled chain reaction in 1942, uranium had already become the object of a concerted U.S. government program to acquire a large supply in anticipation of building a new kind of bomb. Over the years that followed, the government created a freakish market—a total monopsony in which the government was, by law, the only buyer of uranium and could determine supply by setting prices and controlling demand.

The government realized even before the outbreak of World War II in the fall of 1939 that the silvery, dense element could be the key to bombs of magnitude dwarfing anything the world had seen before. One of the first to foresee the implications was Hungarian scientist Leo Szilard, another of the extraordinary European scientific talents who fled Nazi persecution and landed in the United States in the years leading to war and holocaust. Szilard, one of the leading theorists of fission, became concerned about keeping uranium out of Hitler's hands and under free-world control. He shared his concerns with another expatriate Hungarian physicist, Eugene Wigner. Together in the United States in the summer of 1939 and fretting about uranium, they decided to ask Washington to warn the Belgian government of the need to secure its large uranium mine in its Congo colony in Africa. Szilard and Wigner persuaded the most prominent scientific refugee from Nazi oppression, Albert Einstein, to write a letter to President Roosevelt, raising the alarm about uranium.

Investment banker Alexander Sachs, who had access to the White House, agreed to hand-deliver the letter to Roosevelt. By the time Sachs was able to see Roosevelt in the Oval Office and deliver Einstein's letter on October 11, 1939, Germany had already invaded Poland, and the war was on.

The Einstein letter began:
Sir:

Some recent work by E. Fermi and L. Szilard which has been communicated to me in manuscript, leads me to believe that the element uranium may be turned into a new and important source of energy in the immediate future. Certain aspects of the situation which has arisen

147

seem to call for watchfulness and, if necessary, quick action on the part of the administration.

Einstein noted that the United States "has only very poor ores of uranium in moderate quantities," adding that the best resources were to be found in Canada, Czechoslovakia, and the Belgian Congo. The letter suggested that Roosevelt designate someone to coordinate the federal government's interest in uranium, "giving particular attention to the problem of uranium ore in the United States."

Most of the world's uranium in 1939 came from two sources: the Shinkolobwe mine in the Belgian Congo jungle and the former silver mines at Joachimsthal in the Germanic region of Czechoslovakia, annexed as the Sudetenland by the Nazis in 1938. Most geologists believed uranium was an extremely rare mineral. Only a few other locations were known to have appreciable amounts of uranium ore. There was uranium in Canada and some in the U.S. in the Four Corners area where Colorado, Utah, New Mexico, and Arizona meet in the high desert, although many geologists doubted the U.S. prospects.

Uranium had only two uses prior to its development as a source of energy. It was used in mineral form to color glass, including ceramic glazes. Refined, it yielded radium, the rare radioactive element discovered by Marie Curie, which formed as the uranium spontaneously split apart. In the early twentieth century, radium was viewed as a wonder element, described by the *New York Times* in 1908 as having "practically no commercial use, but its value in laboratory experiments has created a demand that cannot be satisfied." The curiously radioactive element became the most expensive substance on earth. The *Times* reported in 1918 that radium was selling for $2.8 million an ounce.

Uranium was generally sought only for the radium. The world's richest source was the Belgian Congo mine under the control of Union Miniere du Haut Katanga, a subsidiary of the mammoth Belgian firm Societe Generale. The ore in the Congo mine was so rich it gave Union Miniere a worldwide radium monopoly.[73] At the time of World War II, Union Miniere's Katanga

---

73 The Katanga region, a large southern portion of the Belgian Congo, is rich in minerals, particularly copper and cobalt, along with uranium. When the Congo won independence

holdings represented over 7 percent of the world's copper production and about 90 percent of the cobalt, an extremely important mineral for the coming war effort.

Journalist Tom Zoeller in his book *Uranium, War, Energy, and the Rock That Shaped the World*, describes how Union Miniere and its director, Edgar Sengier, responded to the accumulating war clouds of the 1930s. "As Europe began sliding toward war in the late 1930s," Zoeller writes, "the market for radium began to suffer, and Sengier closed down the Shinkolobwe mine. He neglected to have it pumped, and the pit flooded with dirty water. A visitor described it as 'a gray ulcer.'"

But there was some unusual interest in the uranium oxide, refined in the Congo from the Shinkolobwe ore,[74] which Sengier had stockpiled by 1937, some five thousand tons in all. Officials from Britain sounded him out rather cryptically about use of the uranium, and never mentioned a price. He accepted an offer from a French delegation for his uranium stockpile, which would give Union Miniere half of the royalties from any patents developed to use the uranium the French scientists were able to develop. Germany's invasion of Belgium killed the nascent deal, and Hitler's forces seized eight hundred tons of uranium sitting on French docks. Some forty-two hundred tons of uranium oxide remained in Katanga.

Sengier, a fierce opponent of the Nazis, concluded that the best way to continue his business following the fall of Belgium in 1940 was to move his operation to New York. He set up business in Manhattan, continuing to control the mines in the Congo, which had not fallen under German hegemony. Sengier had hungry customers for his cobalt, needed in aircraft engines. He also decided to move some of his uranium oxide stockpile from

---

in 1960, the region attempted but failed to secede and split off into a separate country founded on its mineral wealth.

74 Raw uranium ore contains a lot of useless mineral material and varying amounts of uranium oxide (U_3O_8), the first stage in the process that becomes uranium fuel for bombs and power plants. Milling plants that remove the dross and concentrate the uranium oxide, a yellowish powder known throughout the industry as yellowcake, are an integral part of the process. Most mines either have mills associated with them, or the mines contract with milling companies. Most references and usage in this book treat uranium, uranium oxide, and yellowcake interchangeably.

Africa to New York, shipping twelve hundred fifty tons in barrels on two freighters to an empty Archer Daniels Midland warehouse on Long Island.

The United States was beginning to take a serious look at acquiring uranium for its own uses. Einstein's letter to President Roosevelt coincided with new scientific understandings of fission. Articles in the March and April 1939 editions of *Physical Review* confirmed fission, the splitting of the atoms and release of neutrons, in U235. An American Physical Society meeting in Washington at the end of April focused on U235 and how to obtain it from the more abundant U238.

Einstein's letter to Roosevelt got the president's immediate attention. When he finished reading the letter in the Oval Office, Roosevelt said to Alexander Sachs, the intermediary who delivered it, "Alex, what you are after is to see that the Nazis don't blow us up," and then told his trusted aide, Edwin "Pa" Watson, "This requires action."

Roosevelt's instruction resulted, within ten days, in the creation of a government uranium advisory committee including Lyman Briggs, director of the National Bureau of Standards,[75] and a trio of Hungarian expat scientists: Szilard, Wigner, and Edward Teller, invited by Sachs to join the group.[76] The group moved rapidly, reporting to Roosevelt on November 1 that a nuclear chain reaction was possible, although not yet proved, and could be useful for submarine power and might also produce "bombs with a destructiveness vastly greater than anything now known."

Vannevar Bush, a mathematician and electrical engineer who headed the non-profit Carnegie Institution in Washington, soon persuaded his old friend Roosevelt to name him to head a new government group, the National Defense Research Committee. Roosevelt on June 28, 1941, issued Executive Order 8807, establishing the Office of Scientific Research and Development in the executive office of the president, subsuming Bush's research committee. The new organization included a section on uranium, known as S-1. That became the foundation of an Army special project, named the Manhattan Engineering District. The Manhattan Project was born.

---

75 The NBS was the nation's primary science agency at the time, housing almost all of the country's governmental expertise in physics.

76 Einstein was also invited, but did not attend.

One of the first Army officers assigned to the new program was a West Point graduate with a PhD in hydraulic engineering, Kenneth D. "Nick" Nichols. He became the chief "fixer" for the project, locating supplies of needed raw materials and acquiring land for the sprawling enterprise, working for the hard-driving Gen. Leslie Groves.[77] One of Nichols's key tasks was procuring the uranium needed for the bomb builders.

Union Miniere's Sengier, working out of his Manhattan office, approached the U.S. government in early 1942 seeking a market for his uranium, with no success. The creation of the Manhattan Project changed Sengier's prospects. In September 1942, Nichols ventured to Sengier's office and promptly reached a deal to buy the twelve hundred tons of uranium dioxide sitting in the barrels at the Long Island warehouse, and the three thousand tons still in Katanga, at a price of $1.04 per pound. As AEC historian Richard Hewlett recounts, the uranium from the Congo left in "several hundred-ton lots from West African ports. Sea transport over the U-boat-infested waters of the South Atlantic was a dangerous enterprise, but by using sixteen-knot-ships, the company brought through all but two cargoes."

At the same time, Nichols contracted with Eldorado Gold Mines Ltd. for ore from their Canadian mines and for refining Canadian and Congolese ore. He then persuaded Sengier to reopen the Shinkolobwe mine. The Army also contracted with producers on the Colorado Plateau to buy all of their uranium output, only a tiny amount at that time. By 1944, the United States had secured all the uranium it needed for the war effort and effectively controlled the free world's existing uranium production capacity.

But the U.S. supply of uranium at the end of the war was not enough for the needs and ambitions of the newly-created Atomic Energy Commission. The United States had barely enough fissile material on hand to make the bombs that fell on Japan and nothing to spare for new weapons, atomic

---

77 Nichols had a long career in the atomic energy establishment, including serving as the third general manager of the AEC, from November 1, 1953 to April 30, 1955. Nichols later described Groves as "the biggest sonovabitch I've ever met in my life, but also one of the most capable individuals." Clearly, he never worked for Adm. Hyman Rickover. In fact, Rickover technically worked for Nichols when Nichols was AEC general manager, although Rickover never acknowledged any bureaucratic superiors.

bombers, or any of the hyperbolic atomic dreams featured in the popular press and the fantasies of the AEC. The commission and the congressional Joint Committee on Atomic Energy were also convinced that uranium was an extremely scarce mineral. Locking up the supply would require buying up all the uranium the commission could find, at any price needed to secure that supply.

Another issue that troubled the government policymakers was that most of the uranium for the atomic energy program was still coming from foreign sources, mostly from Shinkolobwe and some from Canada. AEC historians Mazuzan and Walker observed, "No proven uranium reserves existed in the United States, and experts doubted that sufficient or even significant quantities of domestic ore could be found."[78] In 1952, the Joint Committee on Atomic Energy's raw materials subcommittee, headed by New Mexico Democratic Sen. Clinton Anderson,[79] commented, "The subcommittee places heavy stress upon the objective of reducing the dependence of the United States upon foreign sources of uranium ore."

The AEC decided that throwing money at the uranium problem would solve it. The commission in April 1948 announced that it would pay a guaranteed minimum of $3.50 per pound for ten years for high-grade U.S. ore. The offer also included a $10,000 bounty for discoveries of new domestic ore bodies.

Initially, the AEC program faltered. Only a tiny amount of new U.S. uranium flowed to the agency. So the commission in March 1951 doubled down, announcing it would extend its purchase guarantee through March 1962. It was simply a matter of the time it takes to find commercial quantities of ore in vast, unexplored territory before prospectors would hit pay dirt. It was quite easy for the AEC to kick off a uranium rush, as most of the land in the Colorado Plateau was owned by the federal government and managed by the Interior Department's Bureau of Land Management. The

---

78 This incorrect assumption colored the U.S. government's uranium policies from the 1950s to the turn of the twenty-first century, leading to actions wildly out of line with geologic reality. Uranium, it turns out, while not as common as dirt, is widely distributed.

79 It should come as no surprise that the JCAE subcommittee was headed by a legislator whose state had a direct interest in the work of the subcommittee— in this case New Mexico, which was a center of uranium mining and milling in the 1950s uranium boom. That's the way the committee operated in every area.

liberal 1872 federal law governing filing of mineral claims on federal land, designed to boost development in the western United States, also aided the twentieth century quest for uranium by making it easy for a prospector to secure a mine site.

MIT economist Paul Joskow, in a famous 1976 paper on uranium markets, described the AEC uranium policies succinctly:

The AEC established a fixed minimum price schedule for the purchase of uranium ore of various qualities and provided firms with additional bonus payments for initial production of uranium, for development expenditures, and for the production of ore with U_3O_8 contents of greater than .20 percent. The AEC also let participation contracts to encourage uranium exploration and paid for access roads to mining areas. The AEC ran the milling part of the supply stream something like a regulated utility. A prospective mill owner would have to apply to the AEC for a certificate of need. If granted, the AEC would sign a long-term (five to seven year) cost plus profit contract for the delivery of a specified quantity of U_3O_8, over the contract period.

The AEC's generous guaranteed purchases and mill subsidies finally produced results. In fiscal year 1948, the AEC spent $35.5 million in "raw and feed materials," out of a total budget of $672 million, or 5.3 percent of the agency's budget. By fiscal year 1953 (unadjusted for inflation), the AEC spent $148 million on raw materials (primarily uranium) from a budget of $1.744 billion, or 8.5 percent. The AEC year-by-year raised the amount of uranium it was buying,[80] from 12,500 tons in 1953 to 15,000 tons in April 1954, upped to 20,000 tons six months later. By February 1956, the AEC quota was 27,000 tons annually. In 1948, the AEC estimated U.S. uranium

---

80 Until 1968, the AEC was the only buyer of uranium in the U.S., so there was no real market and no way of reliably establishing market-based prices for yellowcake. Private companies— primarily electric utilities – were not able to own enriched uranium fuel until August 1964, when President Lyndon Johnson signed the "Private Ownership of Special Nuclear Materials Act," ending the government's 18-year monopoly on uranium. Before then, power companies had to lease reactor fuel from the AEC.

reserves at a million tons. By 1957, according to historians Mazuzan and Walker, the estimate had grown to 70 million tons.

The price the AEC was paying for uranium had also escalated dramatically. The maximum price—which the AEC always paid—rose from fifteen dollars per pound to twenty-five dollars per pound. The extremely generous price provided enormous profits for those who found and processed uranium.

Vernon Pick wasn't the only prospector who got rich from AEC uranium purchases. Mines and mills sprung up all over the Colorado Plateau, along with new car dealers to supply the miners and millers with fancy new trucks; bars, eateries, and extravagant houses popped up in the western desert like spring mushrooms in the forests of the East.

One uranium millionaire, petroleum geologist Charlie Steen, began prospecting for uranium on the plateau in 1949, a hundred years after the famous California gold rush. In July 1952, living in penury and a month after Vernon Pick's pick struck pay dirt, Steen struck it rich southeast of Moab, Utah. The $250,000 mansion he later built for himself and his family to replace their tarpaper shack from his prospecting days was famous for its opulence. The house featured a swimming pool, a lavish greenhouse, and servants' quarters. He named the house *Mi Vida*, the same as his fabulous uranium mine.

Steen lived *la vida loca* in the 1950s. He hosted parties that included Hollywood guests such as Anthony Quinn and Henry Fonda. He flew his private plane to Salt Lake City weekly for rumba lessons. In 1958, the profane and secular Steen persuaded the Mormon majority in the region to send him to the Utah legislature, but he found politics boring and confining, resigning in 1961 in the dying days of the uranium rush.

The boom all too soon came to the predictable bust. Born at the hands of the AEC, the uranium balloon burst as a result of deliberate AEC policy.

17. Naked Shorts at Westinghouse

On October 28, 1957, Jesse Johnson, head of the Atomic Energy Commission's raw materials program, traveled to New York for the annual meeting of the Atomic Industrial Forum, the nuclear industry's trade group. He made an announcement that *Time* magazine two weeks later described as hitting the uranium mining and milling business "with a hydrogen bomb."

Johnson said, "We have reached the point where it is no longer in the interest of the government to expand production of uranium concentrate." After a decade of all-out pursuit of uranium at almost any cost, the AEC was abruptly changing course. *Time* said the AEC "put on the squeeze: any big new uranium discoveries will probably not be able to find a market." In the words of country music icon Willie Nelson, "Turn out the lights; the party's over. They say that all good things must end."

Facing a hostile audience, Johnson explained that the commission now was able to assure a stockpile of fifteen thousand tons of yellowcake a year for the next decade, more than enough for the commission's weapons program and the still unborn commercial electric power business. Johnson said that ending the AEC uranium incentive would force the miners to seek new markets. "Much of this incentive will have to come from confidence in the future market for atomic power," he said.

But the commercial market did not yet exist and was developing extremely slowly. *Time* noted that the nation's first nuclear power plant, which Westinghouse Electric Co. was building for the AEC at Shippingport, outside Pittsburgh, Pennsylvania, was not yet commercially viable. "Westinghouse has spent eighty cents on research and development for every dollar spent on construction," the magazine observed. *Time* added that the Indian Point plant that Consolidated Edison, New York City's utility, was putting up with AEC funds in West Chester County, had seen it costs jump from $55 million to $90 million.

Recapitulating the days of the nineteenth century gold and silver rushes in the Colorado Plateau,[81] the uranium boom of the 1950s turned into the uranium bust of the 1960s. Mines closed, and mills were abandoned, leaving mildly radioactive mill tailings piles behind; once-thriving towns returned to dust. The market the AEC had created was not, in a term loved by environmentalists, sustainable. When the government pulled out its props, the entire enterprise slowly caved in like an abandoned mine.

Among the victims of the AEC–created boom-and-bust was the ebullient Charlie Steen. When the uranium market tanked, Steen attempted various other businesses, investing in an airplane factory, a marble quarry, and a pickle plant. His ventures failed, and the Internal Revenue Service drove him into bankruptcy in 1968, pursuing him for unpaid taxes. In 1971, Steen suffered a severe head injury while prospecting, struck in the head by a wrench attached to a drill pipe. He never fully recovered. Charlie Steen, born in 1919, died of complications from Alzheimer's disease in 2006. His *Mi Vida* hacienda became a Moab steakhouse.

With no government-made market and no commercial demand, the uranium business simply stalled out in the 1960s and then began to crash. MIT economist Joskow wrote, "During this period the uranium industry reached its peak in terms of production and capacity. More uranium oxide was produced by the industry during 1961 and 1962 than has ever been produced subsequently. However, exploration activity peaked in 1957 and then began to decline in response to AEC policy."

Between 1962 and 1968, the uranium industry contracted dramatically, wiping out the small producers and producing mergers and acquisitions among the bigger players with deeper pockets, often integrated oil companies such as Gulf and Standard Oil of Ohio, companies affiliated with international mining conglomerates based in Canada and Australia, such as Rio Tinto Zinc, and foreign government agencies in France, Canada, and South Africa.

---

81 The National Mining Hall of Fame is located in the aptly-named town of Leadville, Colorado. Charlie Steen is among the inductees. Vernon Pick, who got out of mining before the roof collapsed, is not.

The AEC kept in place a program to buy minimal amounts of concentrate at eight dollars per pound, designed to keep some fuel infrastructure in place in case the hoped-for market for commercial nuclear power plants materialized. This was known as the eight-dollar reserve. Beginning in 1962, the AEC began purchasing only five hundred tons of oxide yearly at the reserve price, with purchases running through 1966. The AEC also forbade commercial buyers of uranium—the nascent power industry—to buy foreign uranium from the rapidly expanded supplies coming from French, British, Australian, Canadian, and South African producers. That policy lasted into the late 1970s and helped fuel a secretive international uranium cartel.

The AEC foresaw a boom in nuclear power plants. In its 1962 annual report, the commission predicted five thousand megawatts of nuclear capacity in place by 1970 and forty thousand by 1980.[82] By 1966, it looked like the power plants might finally make their blockbuster appearance in the commercial market. Both General Electric and Westinghouse were heavily promoting their plant designs to utilities, touting the benefits of low and stable fuel prices in order to compete with the dominant coal-fired technology. Anticipating a new market for uranium, uranium exploration and investment in milling capacity began picking up in 1967 in order to be ready to roll by the early 1970s. According to Eugene Grutt, the manager of the AEC's Grand Junction office, "1969 was the peak year with nearly thirty million feet" of exploratory cores drilled.

Both GE and Westinghouse employed uranium fuel as a marketing ploy. They agreed to provide buyers of their reactors with free fuel—usually the first fuel load for the plant, about a year's worth of enriched uranium—and varying amounts of fuel for reloads for up to ten years of operation of the plant. This was a rational sales incentive by the suppliers of the nuclear steam supply systems, because the chief attraction of nuclear power was the prospect of cheap fuel. With the uranium market in free fall, it looked to GE and Westinghouse like supplying fuel to the plant buyers would be a low-cost way to generate market share.

---

82 In 1970, according to U.S. government figures, installed U.S. nuclear capacity totaled 7,004 MW; in 1980, the figure was 51,810 MW.

Pittsburgh-based Westinghouse, with the most experience in power plant technology by virtue of its work in the Navy's submarine program and its development of the reactor for the Shippingport plant, was the most aggressive marketer. For reasons not entirely clear, Westinghouse also made a catastrophic decision about its marketing pledge to supply its plants with uranium fuel far into the future. The company—its excellence concentrated in engineering but not in economics—decided to go short on the uranium market. What's more, Westinghouse decided on what commodity traders call a naked short.

Without getting deeply into the weeds of stock and commodities trading law and theory, a "short," or a "short sale" is a bet by an investor that the price of a stock or a commodity, such as uranium, will fall. The investor buys in the falling market, and recoups by selling when the market price for the stock or commodity eventually rebounds. Investors, particularly in commodity markets, also use shorts to hedge against changes in markets, much as bookmakers "lay off" bets to even out fluctuations and cover themselves regardless of the outcome of the events on which they are taking bets. Smart commodities players and bookies hedge shorts by buying contracts to assure that they are made whole if the market goes up. When an overconfident investor decides to bet against the market and do it without a hedge against disaster, that's a "naked short."

During the sweet spot in the market for nuclear power plants—the late 1960s to mid-1970s—Westinghouse was a major player. By 1975, Westinghouse was the world's largest supplier of nuclear steam supply systems. By 2005, according to the nuclear consulting firm Scientech, forty-eight of the operating commercial nuclear reactors in the United States were Westinghouse pressurized water designs and thirty-five were GE's competing boiling water units.

In making sales to fifteen U.S. electric utility systems in the 1960s and 1970s, Westinghouse took on an obligation to provide its customers with about eighty million tons of yellowcake, which it would enrich and fabricate into finished fuel. Through the early 1970s, Westinghouse covered its marketing offer by buying fuel contracts from uranium suppliers at fixed prices.

In 1972, some genius at Westinghouse decided to get naked, not covering its commitments to its reactor customers with fuel in hand in the form of futures contracts or physical inventory. To a bean counter, this may have looked like a way to save money by not tying up cash in long contracts or physical supply.

But the naked short was a massive bet that the current market for uranium would at least stay the same and never go up. At best, it would continue to fall. As James J. Friedberg, a young lawyer at the Tennessee Valley Authority in the 1970s, wrote later, "Had the market price stayed where it had been for over a decade (in the range of six to eight dollars a pound for U_3O_8), Westinghouse's short sales would not have caused a problem. Westinghouse simply could have bought the uranium on the open market as its needs came due." But if the market went up, Westinghouse was screwed.

At the time Westinghouse decided to strip down to shorts, the uranium market looked very weak indeed. Energy reporter Gene Smith in the *New York Times* wrote in January 1972, "Exploration is down and so are prices. A surplus is building up as nuclear power plants are delayed by environmental regulations. The government plans to unload its stockpile and mining costs are escalating rapidly, particularly because of more rigid safety requirements. 'Personally, I feel sorry for the industry,' George White, president of the Nuclear Exchange Corporation of Palo Alto, California, said in a telephone interview last week."

On the contrary, the market price for uranium started going up in 1973, driven in large part by the flood of orders for nuclear plants from Westinghouse, GE, and other reactor vendors, including Babcock & Wilcox, Combustion Engineering, and General Atomics. Smith of the *Times* wrote in November 1973, "Electric utilities are convinced that nuclear power is the best way to beat the problems of rising costs for fuels and environmental problems. Last week, in separate statements, the chairman of the General Electric Company, which makes nuclear reactors, and the president of the Exxon Corporation, which sells uranium as well as petroleum, agreed that nuclear power plants would be a major factor in electric power generation in this century. But the utilities, needing uranium to fuel their reactors, are finding it more difficult to get firm supply commitments from uranium producers, who want to see their prices increase." Smith reported that Chicago-based Commonwealth Edison sought bids for twenty-two million pounds of uranium for the period from 1981 through 1992. "Glenn W.

Beeman, vice president of the giant Chicago utility, said bids were asked for twenty-two companies 'but not many have been received yet. I guess they're waiting for the market to firm up.'" TVA and General Public Utilities also reported lack of interest on the part of bidders for new uranium supply.

By September 1975, the market price for uranium oxide had hit twenty-six dollars a pound, and rose to over forty dollars a pound through 1980. Westinghouse's bet had turned very sour and extremely dangerous to the energy giant.

Facing the need to eat billions of dollars in its contracts with its customers—an event that was clearly life-threatening to the company—Westinghouse predictably took the dishonorable but easy course of action. The company hinted at what was coming in July 1975, as the *Times* reported on Westinghouse's second-quarter earnings. While the company reported a 75 percent earnings increase for the quarter, compared to the second quarter of 1974, Westinghouse also warned of a dark cloud on the business horizon. "In a separate but related announcement," said the newspaper, "the Pittsburgh-based company also said that it was uncertain about its ability to meet uranium supply demands from its customers beyond 1978 because of a variety of recent developments in the marketplace." Westinghouse linked its problems to the oil price run-up of the first Arab oil embargo of 1973, although oil constituted a small portion of electricity generating capacity in those years.

Westinghouse dropped the radioactive shoe two months later, notifying its customers that it would renege on its contracts. On September 8, 1975, fifteen Westinghouse reactor customers got curt letters informing them that the company would not honor its contracts to supply fuel. Accompanying the letters was a legal analysis by the Chicago law firm of Kirkland and Ellis and an economic analysis by James Lorie of the University of Chicago and Washington economics consultant Celia Gody. The *New York Times* the next day reported that Westinghouse claimed "that the purchase of uranium in the open market at current prices would involve such an unfavorable and large burden as to be 'commercially impracticable' from a legal viewpoint," and the company "is therefore legally excused from a portion of its obligations to deliver uranium." The source of increase in uranium prices, Westinghouse argued, was beyond its control, a function of the Arab oil embargo, "government policies and actions of foreign uranium-producing countries." In short, Westinghouse claimed that the Arabs, the government, and other foreigners made them do it.

The uranium consumers reacted instantly. The *Times* the next day reported that the Tennessee Valley Authority and Richmond-based Virginia Electric and Power Co. were aghast at the Westinghouse move. The TVA, the nation's largest electric power producer, said it took a "dim view" of the uranium supplier's announcement. Vepco senior vice president Stanley Ragone hinted at a lawsuit, saying that his company's lawyers "do not agree that they [Westinghouse] can be legally excused as they contend." The newspaper also noted that Florida Power and Light, also a Westinghouse reactor customer, immediately postponed a $75 million stock offer, citing "uncertainties presented by the Westinghouse announcement."

The implicit threat of legal action quickly became explicit, when the TVA sued the Pittsburgh reactor vendor in federal court on October 20. That was the first of fifteen other lawsuits against Westinghouse: thirteen in federal courts, one in Pennsylvania, and one in Sweden. In its filing in U.S. District Court for the Eastern District of Tennessee, the TVA asserted that Westinghouse had enough uranium on hand to meet its obligations through 1978, but had told the utility it would only get 18 percent of its contract amount of fuel. What soon became the largest civil litigation in U.S. history to date was now underway.

The TVA's status in the suit was quite interesting. No utility in the country went as crazy for nuclear power as the giant regional power agency, which got into the electric generating business in the 1930s, largely as an afterthought. The TVA's original mandate was for flood control along the Tennessee River system and regional economic development. The TVA started generating power because the flood control dams it built on the river system also offered easy opportunities for manufacturing power. The political model for the TVA was the statewide agency that then New York governor Franklin D. Roosevelt created before he moved to Washington, the Power Authority of the State of New York. PASNY had explicit instructions to harness the kinetic energy of the Niagara River.

The TVA had the authority to sell excess power from its hydro network to local distribution utilities, all either owned by local governments, or rural cooperatives owned by the customers. In World War II, driven by the power needs of the war effort—making aluminum and powering the

Manhattan Project—the TVA built coal-fired generating plants to supply the growing electricity load.

Sitting on a major river system in the middle of one of the richest coal fields in the world, nuclear power seemed ill-suited to the TVA's needs. But the longtime head of the agency, Aubrey "Red" Wagner, became convinced of the value of nuclear power in producing cheap electricity, and he plunged the TVA headfirst into the first generation of nuclear power. Between 1966 and 1978, the TVA committed to develop nineteen new base load[83] generating plants— seventeen were nuclear plants. The TVA ordered plants from all of the major nuclear vendors: pressurized water reactors from Westinghouse, Babcock & Wilcox, and Combustion Engineering; boiling water reactors from General Electric; and gas-cooled reactors from General Atomics.[84]

Former TVA lawyer James Friedberg noted that TVA had "special reasons to be indignant" when Westinghouse abrogated its contracts. "Just a year earlier, at an August 1974 meeting," Friedberg wrote, "TVA officials had sought assurance from Westinghouse that the uranium for the initial cores of the Sequoyah and Watts Bar nuclear plants would be delivered. The assurance was sought partly because of rumors of a Westinghouse uranium shortage and partly because of a Westinghouse refusal to offer TVA reload uranium to which TVA officials believed TVA had contractual rights. The Westinghouse officials at the August 1974 meeting gave the TVA personnel the assurance they sought."

---

83 Base load plants are generating units that run all of the time, even when there is little demand for electricity, such as overnight. They provide the base of a utility's power supply. These plants are usually coal-fired, as the cheapest source of electricity. More expensive plants, such as units fired by natural gas, run more often when demand is higher. These are known as "peaking" or "intermediate" units. U.S. nuclear regulators have long demanded that nuclear plants run only in base load mode, rather than ramping up or down to follow demand.

84 Ultimately, only six reactors were built: three GE reactors at Browns Ferry, two Westinghouse units at the Sequoyah site, and one Westinghouse PWR at Watts Bar. A second PWR at Watts Bar went into mothballs in 1985 and construction resumed in 2007, with completion scheduled for 2013. The TVA board in 2011 voted to resume construction on a Babcock & Wilcox PWR at the Bellefonte site in Alabama.

Because the lawsuits against Westinghouse were being filed at federal courthouses all across the country, the federal Judicial Panel on Multidistrict Litigation held a hearing in December 1975. The panel consolidated the cases and ordered that they be tried in the Eastern District of Virginia with Judge Robert R. Merhige in command.

Appointed to the federal court in 1967 by President Johnson, Merhige became a legendary jurist, presiding over a series of crucial school desegregation cases in which he ordered an end to the state of Virginia's policy of "massive resistance" to school integration. In 1970, Merhige ordered the University of Virginia to admit women. He also played a major role in landmark environmental legislation.[85] Merhige was a no-nonsense judge who valued consensus and settlement rather than contention and argument. The *Washington Post* noted in its 2006 obituary, "He was known for his kindness and integrity and for brooking no delays or foolishness in his court, part of the Eastern District of Virginia known as the 'rocket docket.' He once ordered a marshal to remove a man who had fallen asleep in the courtroom. The man, it turned out, was his father."

Merhige convened a pre-trial conference on January 6, 1976, in a venerable Richmond courthouse filled wall-to-wall with lawyers, journalists, and sundry hangers-on. Merhige commented, "I have a feeling that this case is going to be for the legal profession what the Chicago fire was for the construction industry."

The TVA's Friedberg, one of three TVA lawyers involved in the suit, commented, "Discovery entailed the production of millions of documents from all over the world. Many hundreds of days of depositions were conducted, often running simultaneously, at opposite ends of the country or places in between. Over a hundred attorneys spent a substantial portion or all of their time on the case over a three-year period and hundreds more were involved peripherally for a few hours, days, or months."

Westinghouse based its decision to abrogate the contracts and its legal defense on an opinion from its outside lawyers. The lawyers said Westinghouse could justify its actions on the basis of section 2-615 of the

---

85 The University of Richmond School of Law houses the Merhige Center for Environmental Studies, in recognition of his role in settling the lawsuit over the pollution of the James River with the pesticide kepone in the 1970s.

Uniform Commercial Code, a model for business relationships adopted throughout the United States. In a confidential July memo to its client, later disclosed by the *New York Times*, Kirkland and Ellis advised that "delay in delivery or non-delivery in whole or in part by a seller is not a breach of his duty under a contract for sale if performance as agreed has been made impracticable by the occurrence of a contingency the non-occurrence of which was basic assumption on which the contract was made."

The Westinghouse claim drew skeptical reviews from the business bar. Friedberg noted, "Congress has never passed legislation making [the Uniform Commercial Code] generally applicable to federal commercial law. Consequently, government contract law continues to be governed by case law and particular federal statutes, not by the UCC." Alan Schwartz, a business law professor at the University of Indiana, told the *Times*, "If Westinghouse were to lose, it would not astonish anybody. But it may be that the courts will do differently this time because the money is bigger."

It was clear from the start that Westinghouse was on thin legal ice with Kirkland and Ellis's strategy. In his first hearing in January 1976, Merhige demonstrated his penchant for speed and implied he wanted to see the case settled, not tried. "But all of you better recognize that nobody forced anybody to take the case, and you have got it and you are going to have to move on it," he said. "It is as simple as that."

In February 1976, under prodding from Merhige, Westinghouse agreed to parcel out the 14.9 million pounds of uranium it had in its inventory or on order to its utility customers under the terms and prices of the disputed contract. That left Westinghouse still some 60 million pounds short of what it owed to the utilities. The initial settlement also set up a utility committee to negotiate a settlement with Westinghouse and report back to the court monthly on progress.

By October 1978, the case was still pending and Merhige, fed up with the delay, issued a bench order that "Westinghouse did not meet its burden of establishing that it is entitled to excuse from the contractual obligations which the court finds exist with plaintiffs." Merhige then strongly warned Westinghouse and the utilities to settle the dispute. "If anything," he said, "the court is disposed to believe that, just as Westinghouse is not entitled to excuse from its contractual obligations, the plaintiffs are not entitled to anything near the full measure of their prayer for relief…These are cases which I think everyone admits should be settled if at all possible, in the public interest, and they are really business problems, and should be settled as business problems by businessmen, as I have been urging from the very first."

Westinghouse got Merhige's message. The company settled separately with each of its uranium customers over the next two years, agreeing to supply what they could and reimburse the customers for the failed uranium deliveries. The largest settlement was with the TVA, in a May 1979 deal. The TVA said it received $130 million in cash, goods, and services, while Westinghouse reported the settlement cost it $36 million. As Friedberg explains, the difference, "other than each side telling its own fish story," was because there were elements that Westinghouse could provide to the TVA with little cost to the vendor but were worth real money to the utility. Also, the TVA got some mineral rights that were close to other reserves it already owned in Wyoming. Because of the location of the uranium reserves, they were worth more to the TVA than to Westinghouse.

The total costs of the settlement took a big chunk out of the Westinghouse bottom line. The company reported that it had a net loss of $74 million for 1979, most of it related to the uranium settlements. By contrast, the company had recorded 1978 profits of $243 million. For the fourth quarter of 1979, when the settlements really started biting, Westinghouse booked a loss of $117 million. Those figures represented settlements of thirteen of the sixteen outstanding uranium disputes. Westinghouse also said it had put aside a $118 million reserve for future uranium settlements.

The same day it released its 1979 earnings, Westinghouse said it had reached a settlement with Union Electric in St. Louis. Westinghouse chairman Robert E. Kirby said the Union Electric deal would result in a pretax loss of $125 million, consisting of $55 million in cash up front and another $55 million if Union Electric finished its second nuclear unit, scheduled for completion in 1987 (and later cancelled).[86]

Westinghouse ended up buying the most expensive shorts in U.S. business history.

---

86 The balance of the $125 million would have come from what Westinghouse might get from a suit it had filed against international uranium producers.

18. The Great Uranium Conspiracy

It was an unremarkable day in Sacramento, California, when Jim Harding arrived at his office at the California Energy Commission in the summer of 1976. Harding, a young resource economist who had previously worked for the feisty environmental advocacy group Friends of the Earth, was an assistant to Ronald Doctor, an economist who served on the commission created by the state legislature in 1974 to serve as an antidote to the perceived industry-bias of the California Public Utilities Commission.

As Harding recalled thirty-five years later (unsure of the exact date of the transaction), he was in his office that morning when Dale Bridenbaugh, a nuclear engineer who had recently left General Electric with two other engineers in an anti-nuclear blast against their former employer, showed up with a two-inch-thick package wrapped in plain paper. Bridenbaugh had no idea what was in the package. He was delivering it to Harding following a visit to Australia. In Australia, the local chapter of Friends of the Earth asked Bridenbaugh to take the package to Harding, someone they trusted who was in a position where he could make use of the information.

Harding said later he didn't know who in the Australian group decided to ship the papers to him. "I never really knew who in Australia had gotten it to me," he said in an interview. "I had done a speaking tour there on uranium, so that's probably how I came to their attention."

When Harding peeled off the wrapper, he discovered a trove of material from the files of Mary Kathleen Uranium, the only Australian uranium mining company at the time. The pages documented two years in the history of a world-wide government and industry cartel organized to manipulate the price of yellowcake on the world market. The Mary Kathleen papers kicked off a major political and foreign policy dispute involving virtually all of the uranium producers and users around the globe and straining relationships among the United States and some of its closest allies, including France, England, Australia, and Canada.

As it turned out, the uranium cartel was a direct product of the misguided policies and practices of the Atomic Energy Commission with regard to the supply of uranium. By first creating a glut of uranium in order to supply its burgeoning weapons program, the AEC produced a predictable result: a crash in prices that destroyed all but the biggest U.S. players in the market. Then, reacting awkwardly to the unintended consequences of its push for production at any price, the AEC furthered the reach of its government-created buyer's monopoly. The agency continued to deny potential customers access to the uranium market on their own, forcing them to transact their business through the U.S. government. Finally, the AEC effectively denied foreign producers access to the U.S. uranium market—the largest in the world—by erecting protectionist barriers to foreign suppliers.[87]

The foreign producers—governments themselves plus the multinational, trans-governmental firm Rio Tinto Zinc—responded in the only way they thought they could. They created a marketing cartel designed to increase prices outside the restricted U.S. market and divvied up the market among the members of the cartel, which they dubbed "The Club."[88] That these actions violated U.S. law mattered not at all to the international cartel. They viewed the United States as a commercial enemy and U.S. antitrust law as a cynical device to restrict competition in uranium markets. One history of the Club noted that the U.S. embargo "created bitter feelings among foreign producers who were also in desperate straits, and it divided the industry into two parts—the United States and the rest."

---

87 The AEC policy prohibited enrichment of foreign-supplied yellowcake to make it useful as reactor fuel in U.S. enrichment plants for use in U.S. reactors. At the time, the United States had the only uranium enrichment plants in the world, which had been created for the Manhattan Project.

88 The Club soon decided that, for public relations purposes, it needed a more anodyne name, so it followed the pattern set by Rio Tinto Zinc after World War I and voted to call itself the Uranium Institute, with headquarters in London. After the cartel collapsed, the Uranium Institute continued, and in the early days of the twenty-first Century became the World Nuclear Association.

This attitude has persisted well into this century. In a 2004 memoir, Terry Price, a well-known British figure in the world of uranium and key staff player in the uranium cartel, commented that the purpose of the Club "was to defend the embryonic nuclear industry against the deliberate exploitation of massive purchasing power by a monopsonistic uranium customer. Once the whole picture was revealed, the Club appeared to be one of the world's more forgivable cartels."

When Harding opened the package from Australia in August 1976, he knew he had something hot. There had been rumors of the cartel in the trade press since the formation of the Club in 1972. *Nuclear Industry* magazine in its February 1972 issue reported on the formation of the cartel at its first meeting in Paris: "Despite persistent rumors that the major uranium producing countries—except the United States—would soon try to end the price cutting that has been hurting them all, the news that they are actually trying to negotiate an armistice surprised most observers. The first word of the move was an Australian government announcement that an official delegation would go to a Paris meeting with French, South African, and Canadian delegations to discuss future agreements to secure orderly marketing." The *Wall Street Journal* and the *Journal of Commerce* both reported on the birth of the cartel in Paris February 1–4, 1972.

Westinghouse chairman and CEO Robert Kirby, who took charge of the company in mid-1975 and led it during the uranium business meltdown, shortly before he took office noticed what he thought was international hanky-panky. In a history of Westinghouse, the *Pittsburgh Post-Gazette* newspaper wrote, "Early on in his uranium work, Kirby began casting a suspicious eye across the Golden Triangle to Gulf Oil's art-deco headquarters. Gulf was a uranium supplier, and Kirby smelled a rat."

By mid-1975, the *New York Times* reported that "the Justice Department is investigating charges that a new international uranium cartel violates United States antitrust laws." In fact, a federal grand jury in Washington had been meeting for several months investigating the existence of a uranium cartel and its relationship to U.S. antitrust law.

Harding shared his trove of documents with his boss, Ron Doctor. Together, they contacted Leonard Ross, a brilliant young lawyer whom Gov. Jerry Brown (in his first term as chief executive of the Golden State) had appointed to the public utilities commission.[89] At Ross's insistence, the California officials contacted the Justice Department, which expressed an interest in their documents. "When we contacted the DOJ," Harding recalled, "they were instantly interested. They told us to make two copies, put one to them in overnight mail, and send the other by regular mail, and keep the original ourselves. They also then sent a courier for a copy. They obviously wanted it."

The package they mailed to Attorney General Edward Levi and Senate Foreign Relations Committee chairman Frank Church (D-ID) included a cover letter. In it, Richard Maullin, energy commission chairman, and commission members Emilio Varanini, Doctor, Ross, and Harding wrote, "If the foreign cartel continues and America's nuclear commitment increases as planned we will be at the mercy of a uranium OPEC."

The Mary Kathleen papers provided incontrovertible evidence about the cartel, in fascinating detail. It outlined how the members of the Club divvied up the market, set prices, determined who would win in allegedly competitive auctions, set up mechanisms to discipline its members, and maintained secrecy. As the documents showed, the kingpin of the organization, and its largest beneficiary, was Rio Tinto, with operations in all of the countries represented in the organization. The Club's members included the governments of France, Australia, Canada, and South Africa, although the governments' roles were indirect, working largely through private or government-owned uranium companies, such as Gulf Minerals Ltd. Canada. The governments operated sub rosa. Only those with a "need to know" were fully apprised of the activities.

The Club tried to maintain strict secrecy for a number of reasons, including the wish to avoid political controversy, which occurred in Canada once the Mary Kathleen papers surfaced. The cartel also wanted to fly under the radar of various national antitrust laws, particularly the very strict U.S. antitrust regime. Perhaps most importantly, the Club did not want to overtly confront the U.S. Atomic Energy Commission, which could withhold its enrichment services, necessary to turn uranium yellowcake

---

89 Harding, Doctor, and Ross all had close ties to California-based Friends of the Earth, parent of the Australian group of the same name that received the stolen Mary Kathleen files on the Club's activities.

into useful reactor fuel. At that point, the United States enjoyed a virtual monopoly on commercial uranium enrichment services. The history of the cartel, *Yellowcake*, by June H. Taylor and Michael D. Yokell, commented, "The problem of a possible refusal by the United States AEC to enrich Club uranium arose over the cartel's policy of discriminating against middlemen." The most significant middleman in the uranium market at the time was Westinghouse.

The four producing nations in the cartel excluded their domestic markets from the price-fixing arrangements, and the cartel deliberately avoided the U.S. market. The exclusion of the United States was not only a prudent precaution but reflected the practical state of the market. Given the U.S. ban on uranium imports, the foreign suppliers had no direct access to the U.S. market. The Club believed it was protected from troubles in the United States, but that protection proved to be weak.

The Mary Kathleen papers exposed the Club, creating the publicity and the political problems the cartel hoped to avoid. The immediate response of the governments involved in the cartel was to deny the accusations. A spokesman for the French government said his country had "no knowledge of any such cartel in which French interests, public or private, would be represented." Questioned about the 1972 meeting in Paris, the French response was that the meeting was open and "even distributed a final communiqué in French and English explaining its purposes." The United States was not invited to the meeting, said the spokesman for France, because the United States was not a uranium importer and could not be hurt by the actions of the Uranium Institute.

But when the Australian uranium archive came to light, Westinghouse immediately filed an antitrust lawsuit against twenty-nine uranium producers, including Gulf, as a counter to the utility suits against Westinghouse for its failure to provide contracted fuel. In a suit filed in October 1976, Westinghouse claimed that the Club pushed up uranium prices by 500 percent, leading to the collapse of Westinghouse's short position in the uranium market. The suit filed in U.S. District Court in Chicago sought triple damages from U.S. companies including Gulf, Engelhard Minerals, Anaconda (acquired by oil company Atlantic Richfield), Getty Oil, and Utah International (acquired by General Electric). The first step in the lawsuit was discovery, where Westinghouse began sorting through the business records of the defendants, notably those of its Pittsburgh neighbor, Gulf.

At the same time, Congress was getting into the act. The House Interstate and Foreign Commerce Committee's oversight and investigations subcommittee in November 1976 began a series of probes of the uranium market, under the leadership of California Democrat John E. Moss, ironically representing a Sacramento district. Moss, the son of a coal miner, was a committed New Deal liberal and an aggressive investigator who served thirteen terms in the House from 1953 to 1978 and was the father of the Freedom of Information Act.[90] The Moss subcommittee's first hearing took place in Sacramento.

In May 1977, Moss's staff picked up some hints, from Westinghouse's trolling through the Gulf papers during discovery, that Gulf was intimately involved in the cartel's price fixing. If true, that would be a clear violation of U.S. law and a legal hook on which to hang a high-profile congressional investigation. The problem with the story of the uranium cartel was that tying it to the U.S. market and U.S. law was proving difficult.

The Moss subcommittee scheduled a hearing for early May 1977 and subpoenaed Westinghouse to appear and produce all the material it had obtained from Gulf during legal discovery. On the morning of the hearing, May 2, 1977, Gulf went to court, getting a temporary restraining order from a federal court in Washington preventing Westinghouse from turning over the discovered documents. Westinghouse, placed in what one historian called a "no-win position," chose to honor the court order. In response, congressional lawyers threatened a contempt of Congress citation—arguing that settled law holds that Congress, as an independent body under the U.S. Constitution, is "immune from judicial interference and has a right, independent of any other branch of government, to obtain information for which it issues a subpoena." When Westinghouse continued to follow the court's restraining order, Rep. Henry Waxman (D-CA), acting as chairman of the full Commerce Committee, found Westinghouse in contempt.

---

90 Moss served on the Joint Committee on Atomic Energy from December 1974 through 1976. He was initially appointed to replace California Democrat Chet Holifield, who resigned.

A day later, U.S. District Court Judge George Hart removed Westinghouse from the horns of its legal dilemma, refusing to turn the temporary restraining order into an injunction against release of the Gulf material to the congressional committee. Westinghouse lawyers immediately turned the material over to the Moss subcommittee.

Still fighting to hide its role in the cartel from public scrutiny, Gulf tried to persuade the congressional investigators to withhold the material from the public. The oil company's lawyers offered two reasons why the material should be kept confidential, neither of which convinced the investigators. First, said Gulf, the papers were protected by the lawyer-client privilege against public disclosure, a common-law precedent. Also, Gulf said, release would violate Canadian law designed to protect information about uranium supplies.[91] The committee concluded that the privacy privilege didn't apply when the advice of the lawyers was how to commit fraud, and that Canadian law doesn't apply to the U.S. Congress, particularly when it appeared that Gulf helped formulate the Canadian regulations in order to prevent exposure in the United States.

The committee also issued subpoenas for various foreign participants in the cartel to appear at the hearings. The governments and private business officials largely ignored those orders, although it sometimes cramped their travel plans. In *Yellowcake*, Taylor and Yokell recounted a fiasco in which the Moss investigators learned that an executive from a Canadian uranium firm, Rio Algom, was at a U.S. airport and notified U.S. Customs to pick him up. But the executive, G.R. Albino, made his flight before the feds could nab him.

As the hearings got rolling in June, Canada launched a public relations counter-offensive. On June 16, 1977, Canadian finance minister Donald S. Macdonald, speaking in Ottawa, attacked the Moss investigation and the U.S. government. He called on President Carter, who

---

91 In September 1976, as the information from the Mary Kathleen papers became public, Canadian prime minister Pierre Trudeau and his cabinet adopted a regulation—called an "order in council"—making it a crime for a Canadian to even discuss, let alone release to the public, information about the cartel. Trudeau later acknowledged that the order was specifically aimed at preventing release of information to American officials.

took office in January, to halt the investigation. Macdonald, previously Canada's minister of energy, mines, and resources, said, "We acted to protect ourselves from these predatory American tactics, and now they are saying 'you are maintaining a cartel.'" Macdonald added, "They have an ambassador here in Ottawa. Maybe he had better take a message back to Washington that there is not one law for the United States and a different one for everybody else."

The next day, Macdonald, never denying the existence of the cartel or his government's support for it, told the Canadian parliament that American policies in the early 1970s, including the export ban to protect U.S. producers, forced the creation of the international cartel. Macdonald argued to the House of Commons that the United States was trying to impose its antitrust laws on Canada. "I don't regard that as a friendly act," he said.

Newly appointed U.S. Attorney General Griffin Bell, in Canada for bilateral talks on legal issues between the countries, and U.S. ambassador to Canada Thomas Enders were sitting in the gallery during Macdonald's attack. Reuters reported that Bell said later that the United States wanted to settle its dispute with Canada through "conciliation and accommodation," but without elaboration.

As Macdonald was talking to reporters in Ottawa, the Moss subcommittee opened its hearings in Washington on the uranium cartel by grilling Gulf Oil Corp. executives about the meaning and intent of the documents the committee released during the hearing. Gulf chairman and CEO Jerry McAfee and S.A. Zagnoli, president of Gulf Minerals Resources Corp., the uranium subsidiary, somewhat contradicted the position the company had laid out in a prepared statement. McAfee and Zagnoli admitted to the subcommittee that the cartel accomplished its goal of raising uranium prices in the international market, if only at the margins. Zagnoli testified that, in his view, the cartel was "moderately successful" in raising international uranium prices. In his written statement, undoubtedly prepared earlier by Gulf's lawyers, McAfee argued, "The foreign uranium cartel, which was the offspring of the Canadian government and the government-owned uranium industries of other countries, simply had no discernible impact on either the domestic or foreign uranium commerce of the United States."

174

A young, aggressive freshman Democrat from Tennessee, Albert Gore Jr.,[92] pushed Zagnoli to admit that the cartel pushed up uranium prices in the United States. Gore was attempting to buttress the case of the Tennessee Valley Authority, which had joined Westinghouse in its suit against Gulf, its unfriendly neighbor and adversary in the multi-million-dollar uranium supply lawsuit. The most that Zagnoli would concede was that, if there was an effect on U.S. prices, it was "insignificant."

But the Gulf documents did not produce the kind of self-incriminating evidence the congressional investigators had hoped for. As the *New York Times* coverage indicated, the record was inconclusive. Washington bureau investigative reporter David Burnham reported that some of the documents "indicated that Gulf had taken part in the uranium cartel as a result of pressure from the Canadian government and that the marketing agreements had not influenced the price of uranium in the United States." A June 1972 legal memo by Gulf lawyer Roy D. Jackson advised the company, "The fountainhead of our antitrust defense is the effective Canadian government direction that Gulf participate in the cartel, buttressed by the project minimal impact on the trade or commerce of the United States." Jackson's legal memo also argued that it was important for Gulf "to become a sophisticated and substantial participant in worldwide uranium matters as it was for us to undertake similar efforts with respect to oil and gas thirty or forty years ago. Having a representative on the governing board of the world marketing organization would be helpful in achieving this objective."

By the end of the summer, the Moss investigation had fizzled out. Given the inconclusive records from Gulf, the apparent attempts by the cartel to avoid an entanglement with the U.S. market and antitrust laws, and pressure from foreign governments on the Carter administration, Moss gave up. The subcommittee held a final round of hearings in Nashville, at Gore's urging. The August 15, 1977, hearing was designed to give the

---

92 Gore, whose father was U.S. senator from Tennessee from 1953–1970, was first elected to the House of Representatives in a 1976 special election, with 33 percent of the vote in a multi-candidate field. He was twenty-eight. Gore was elected to the U.S. Senate in 1984, and became the forty-fifth vice president of the United States in 1993, serving under Bill Clinton. Gore was the Democratic nominee for president in 2000, losing to George W. Bush, although Gore won the popular vote (51 million votes for Gore to 50.5 million for Bush).

TVA a podium to make its charges against the cartel, but the power agency's appearance disappointed Gore, who presided at the session, and others seeking to establish Gulf's perfidy. Red Wagner, the TVA's chairman, was unable to provide definitive evidence that the cartel pushed up the price it was paying for uranium fuel. The TVA's Jack Gilleland, assistant power manager, told the subcommittee that the power agency had negotiated with cartel members for twenty million pounds of uranium without ever raising the issue of the cartel and controlled prices. It also appeared that the cartel had disbanded its price-setting regime, although the Uranium Institute remained as a trade association.

The *New York Times* headline captured the subcommittee's frustration: "Effect of Uranium Cartel on Electricity Eludes Inquiry." The article's first paragraph said, "A Congressional subcommittee investigating an international uranium cartel, which says it has disbanded, failed today in its efforts to determine what impact the price-fixing plan has had upon domestic electricity prices."

Over the next few months, the lawsuits brought by Westinghouse and the TVA against nineteen U.S. uranium producers predictably petered out, as plaintiffs and defendants worked out negotiated settlements. Cloaked by confidentiality agreements, the details of the settlements with Westinghouse were never revealed. But the TVA, as a federal government entity, had to reveal its deals, which were most likely in line with those between Westinghouse and the producers. According to Friedburg, "Cash payments to [the] TVA have been relatively small (a few million dollars compared to potential damage claims of hundreds of millions)."[93]

The Justice Department's antitrust investigation of the uranium cartel also sputtered to a close. The DOJ staff concluded that the cartel had broken U.S. law, but the Carter administration refused to take the case any further. Justice sought no indictments. The only legal action was a 1978 Justice Department misdemeanor charge against Gulf, with a penalty of a puny $50,000. Gulf copped a nolo contendere plea, agreeing to the fine but not admitting any guilt in the matter.

---

93 One settlement was potentially more significant, although it involved no cash. Canada's Rio Algom agreed to nullify its contract with the TVA, made under cartel conditions, to supply the power agency with seventeen million tons of uranium.

Did the cartel push up U.S. uranium prices? Probably not—or not to any significant degree. While U.S. uranium prices rose from under six dollars per pound to over twenty dollars by the mid-1970s, that surely had more to do with the freeing of the market from control by the Atomic Energy Commission. As part of the policy decision to break up the U.S. government's postwar atomic energy octopus, dividing the regulators from the atomic power promoters and ending the congressional Joint Committee on Atomic Energy, the government also let free-market forces have their way with uranium supply, and the artificial prices of the AEC regime ended. Had the AEC not decided to run uranium as a government enterprise from the start—a failed experiment in socialism—the cartel likely would never have come to life.

19. Breeding at the Turkey Farm

Tom Swift, the fictional teenage scientist and prodigious inventor, would surely have found a kindred spirit some forty years later in David Hahn, a real teenager who turned his pursuit of a Boy Scouts atomic energy merit badge into a tiny, crude but dangerous, backyard breeder reactor. It truly was life unintentionally imitating art, if one dare call the Tom Swift series of books for boys "art."

Hahn, who grew up in Commerce Township in central Michigan, was both an "archetypal suburban boy" and a scientific prodigy, turning his bedroom into a hazardous chemical waste dump by the time he became a teenager in 1989. His boyhood experiments with chemistry sets and household cleansers and other products yielded dyes, caustics, and explosives. One account of his activities described his boyhood bedroom: "The walls were badly pockmarked from a multitude of chemical explosions, and the carpet was so stained that it eventually had to be ripped out. Even the padding and plywood subflooring underneath was stained blue from spills of indole, an alkaloid derived from indigo pigment that David used to make natural-highlight shampoos."

Then David discovered the atom. His father Ken, a former Boy Scout, pushed his son into joining the scouts as a way to harness his energy and inquisitiveness. David loved the scouts and began assembling an impressive array of merit badges, the path toward the ultimate in scouting status—an Eagle Scout. David soon became enamored of the atomic energy badge, which the scouts had offered since 1963, when then AEC chairman Glenn Seaborg worked with the Boy Scouts of America to create the badge.[94]

David plunged head first into the study of atomic energy, focusing on several of the options for fulfilling the merit badge requirements, including "build a Geiger counter" and "build a model of a reactor." David earned the badge by 1991, when he was fifteen, but that was only the beginning of his interests in the atom. Hahn set out to build a working model of a breeder reactor, one that would turn thorium-232 into fissionable uranium-233. That task consumed him over the next four years.

It isn't known whether his crude atomic energy reactor actually accomplished transmutation, using thorium from the mantles of camp stoves, radium from old glow-in-the-dark camp stoves, aluminum foil, and expertise gathered by posing as a physics professor in letters to the U.S. Nuclear Regulatory Commission. But he did leave a substantial radioactive mess, contaminating a garden shed and yard. Federal officials from the NRC and the Environmental Protection Agency in radiation moon suits had to mount an atomic cleanup with a price tag (paid by taxpayers) in excess of $40,000.[95]

What Hahn tried to do was what physicists and engineers in the United States (and elsewhere) had been trying for decades, with only slightly more definitive success: to create a process that could use the power of the fission process to produce more fuel than it consumed. Called breeding, if it could work at an economic scale, breeding could render moot the question of the supply of uranium in the world. Breeding came to be the Holy Grail of atomic energy, with little to show in results. But the lack of progress did not deter the true believers in the slightest.

Nuclear physicists discovered the principles of breeding early in their inquiries into fission. In 1939, Danish physics pioneer Niels Bohr theorized in the journal *Physical Review* on fissionable elements beyond uranium. In

---

94 In 2005, the Boy Scouts of America changed the name of the merit badge, the 104th of its badges, to the "Nuclear Science Merit Badge" and in January 2011 greatly modified the requirements to earn the badge to focus on radiation safety, not a factor in the earlier requirements.

95 A professionally produced short documentary film of David Hahn's atomic escapades, made in 2005, is available on this books's web site at www.toodumb.org.

1941, chemist Glenn Seaborg discovered how neutrons bombarding U-238 could produce plutonium, a fissionable element. His team followed with the discovery of how to make fissionable U-233 from thorium-232, a very abundant element that is slightly radioactive but will not sustain a fission chain reaction.

The advantage of a breeder was obvious—the ability to create limitless fuel for civilian power plants and the explosive cores of atomic bombs. In April 1944, a group of the atomic illuminati, including Fermi, Szilard, Wigner, and Alvin Weinberg, met in Chicago to talk about how to use fission to light lights, power industry, heat homes, and breed more fuel than the reactions used. In 1945, Fermi proclaimed, "The country which first develops a breeder reactor will have a great competitive advantage in atomic energy." But breeder technology had to wait for the bomb-building Manhattan Project to come to an end before it got full government attention.

One of Fermi's students and closest Chicago associates, Walter Zinn, had developed an expertise in reactor design and construction. Zinn was the scientist who actually pulled the control rod out of Fermi's pile in the squash court under the football stadium at the University of Chicago on December 2, 1942, starting the world's first sustained chain reaction.

By about the same time as the 1944 meeting, Zinn had developed a theoretical design for a plant that would generate both electricity and excess fuel. Zinn's reactor would use fast neutrons—atomic particles not slowed by a moderator in order to improve the chain reaction in U-235—to breed plutonium-239 from natural uranium-238. One important element in Zinn's thinking about breeders, which would bedevil all future breeder projects, was the need for a way to cool the reactor that did not also slow down the neutrons. Most conventional power reactors use ordinary water to both cool the machine and slow the neutrons.

By early 1946, Zinn's conceptual breeder design matured. He would use a mixture of explosive liquid metals—sodium and potassium—as a reactor coolant. The fuel would be bomb-grade U-235[96] encased in aluminum tubes, with the whole assembly surrounded by a cylinder of U-238. Zinn figured that the neutrons from the fission of U-235 would bombard

---

96 The fuel for EBR-I was made from uranium enriched to 94 percent U-235.

the natural U-238 blanket, creating more plutonium than the highly-enriched uranium fuel that the reactor burned up.

In 1947, the AEC approved construction of Zinn's Experimental Breeder Reactor–I, which came to be colloquially known as Zinn's Infernal Pile. Believing the reactor potentially too dangerous to build at Argonne, near Chicago, Zinn pushed the AEC to find a remote site. In 1948 the commission came up with a former Navy ordinance proving ground in the high, remote eastern Idaho desert near Arco. At the time, the scientists and engineers called it Argonne West. The site soon would become the nation's reactor test site, originally given the name National Reactor Testing Station (NRTS), and today known as the Idaho National Laboratory. Zinn estimated his experimental breeder would cost $2.6 million and need forty kilograms (about eighty-eight pounds) of the AEC's very sparse inventory of highly-enriched U-235.

Construction began in Idaho in 1949. A Chicago crew arrived in early 1951, as the reactor building was being completed, to install the core. In May, Zinn tried to operate the reactor, but it proved a dud. There wasn't enough fuel in the football-sized core to sustain a reaction. It took another three months to get more highly-enriched uranium from the AEC stockpile and manufacture beefed-up fuel rods. In August 1951, the plant went critical and began four months of low-power tests.

On December 21, 1951, Zinn's team brought the reactor up to full power. Steam raised by the hot reactor coolant ran to a small steam turbine generator. The most enduring image of EBR-I resulted: four light bulbs glowing brightly from electricity produced from atomic power—the first in the world. The next day, the reactor was generating enough electricity for the entire building.

But the real mission of EBR-I wasn't to generate electricity. It was to prove that the machine could make excess fuel. In June 1953, the AEC announced that the machine, with a power output of 1.2 MW in heat (0.2 MW of electrical output, about the same generating capacity of a small, gasoline-powered generator) was making one new atom of Pu-239 for each atom of U-235 it was burning.[97]

---

97 The technical term the scientists and engineers use is breeding ratio, which is the ratio of uranium burned to plutonium produced. A ratio of 1.0, which EBR-I demonstrated in 1953, is break-even. Eventually, upgrades at EBR-I showed a breeding ratio of 1.27.

While it could breed, EBR-I had inherent problems. The first was what the physicists call a "prompt positive power coefficient of reactivity." In ordinary language, this means that as the power increases, the nuclear reaction speeds up. It is a positive feedback that can lead to an out-of-control reactor. The core could melt and collapse.

The scientists at Argonne understood this problem and used EBR-I to examine the instability of the technology. In late November 1955, during an experiment on the positive reactor feedback, some 40–50 percent of the EBR-I core melted. The Argonne West team removed the damaged core—very carefully—and repaired and operated tiny EBR-I until the end of 1963.[98] A second reactor—EBR-II—operated as a breeder until 1969.

The Arco breeders established the parameters of future U.S. breeder reactor designs. The machines would be called LMFBRs, standing for "liquid metal fast breeder reactors." "Fast" refers to neutrons, which are not slowed down as they are in conventional fission reactors. The faster the neutrons, the more plutonium will result from the reaction. "Liquid metal" refers to the coolant, the metal sodium. Liquid sodium is great at capturing and transferring heat. Sodium also boils at a very high temperature; unlike water, it doesn't have to be under pressure to keep from boiling away as a gas. Finally, sodium, unlike water, doesn't corrode the metal pipes, pumps, and valves it runs through.

But sodium has real problems, which ultimately made it the chemical Achilles' heel of breeder reactors. Sodium, with its high melting temperature has to be kept hot at all times, or it turns solid. Sodium is also extremely nasty when it contacts water, including water vapor. It catches on fire and makes dense clouds of billowing white smoke. Sodium also absorbs neutrons, producing radioactive Na-24 as it passes through the reactor core. So the liquid metal coolant must be kept separate from the liquid sodium used to generate steam in heat transfer equipment known as steam generators. All LMFBRs have two liquid sodium loops: one to cool the reactor and the other to make steam.

---

98 EBR-I is now a Registered Historical Landmark and is open to the public every day from Memorial Day weekend through Labor Day from nine to five. It is located off U.S. Highway 20/26, eighteen miles east of Arco, Idaho.

Even before EBR-1 came to life, an electric company executive had become sold on the promise of the breeder and was assembling a group of like-minded businessmen to launch a commercial breeder. In December 1951, Walker Cisler, a brilliant, Cornell-educated engineer,[99] was promoted to president of Detroit Edison, a large, investor-owned utility serving the Motor City and surrounding Michigan suburbs. Cisler, who had been Edison's chief engineer, ran the utilities program for the Supreme Headquarters Allied Expeditionary Force during World War II, helping rebuild worn-torn electric systems in Europe.

Returning to Detroit in 1945, Cisler became an advocate of the future of atomic power for generating electricity. As executive secretary of the AEC's Industrial Advisory Group in 1947–1948, he advised the commission on its policy for civilian electricity generating plants. Historians George Mazuzan and Samuel Walker described Cisler: "Handsome, articulate, pragmatic, and confident, he moved with ease among top government officials as well as in the industrial, social, and civic circles of Detroit."

Cisler believed that the best way to develop nuclear electric generating plants was through the private sector. He was also convinced that building a commercial-scale breeder reactor was a reasonable and prudent investment in the future, not, as many experts believed at the time, a wild shot in the nuclear dark.

When he was named Detroit Edison president, Cisler, along with officials from Dow Chemical of Midland, Michigan, presented a plan to the AEC to develop a commercial-scale breeder that would generate 100 MW of electricity and produce its own fuel after its initial fueling with highly-enriched uranium from the AEC stockpile. The AEC approved the design phase of Cisler's project, to be known as the Enrico Fermi Breeder Reactor Project, on December 19, 1951, two days before Walter Zinn powered up EBR-I.

Cisler began raising private-sector investment for his project. By October 1952, he had formed a nuclear development department at Detroit Edison and lined up fifteen investor-owned electric utility companies to support his breeder, along with an array of manufacturing and engineering companies. They formed a technical group, with Cisler as chief and

---

99 Cisler was a founding member of the National Academy of Engineering.

legendary physicist Hans Bethe as a consultant, to begin detailed reactor design.[100]

In 1954, Congress rewrote the 1946 Atomic Energy Act to boost civilian nuclear generating plants, following Eisenhower's Atoms for Peace initiative. The AEC announced it would accept proposals from the private sector for reactors under its power demonstration program. In 1955, Cisler formed the Power Reactor Development Company (PRDC), which he headed, to build, own, and operate a 66 MW breeder reactor on nine hundred acres at Lagoona Beach on the shore of Lake Michigan between Detroit and Toledo, Ohio.

Under AEC procedures, the Fermi project was subject to review by the Advisory Committee on Reactor Safeguards. The ACRS established a subcommittee, chaired by Harvey Brooks, engineering and applied physics professor at Harvard University, to review the plan. Ultimately, the ACRS was quite skeptical of Cisler's leap of engineering faith, advising more research—including work addressing serious safety questions about an untested technology, such as the positive feedback that could cause the reactor to melt down. PRDC, said the advisory committee, was advocating a project so bold "as to risk the health and safety of the public."

Nonetheless, PRDC formally applied for an AEC construction permit in January 1956, and the commission approved the project on August 4, 1956, brushing aside the ACRS objections and stressing that it would not issue a final operating license until the safety questions were fully resolved to its satisfaction.[101] PRDC broke ground four days later and started pouring concrete in December. After a legal detour through the U.S. Supreme Court on a challenge to the AEC procedures by four labor unions, which

---

100 Dow withdrew from the consortium, called Atomic Power Development Associates, and later hooked up with Consumers Power, another investor-owned utility, headquartered in Midland, to develop a failed nuclear plant that eventually ended up being converted to run on natural gas.

101 The unusual dispute between the advisory committee, which had no legal authority in the AEC, and the commission and its staff shocked important congressional leaders, who felt the commission had acted precipitously. As a result, Congress in 1957 changed the Atomic Energy Act to make the ACRS a statutory body with a responsibility to review every license application before final AEC action, and to make those reviews public.

the AEC won, the AEC granted a conditional operating license and Fermi I went critical on August 23, 1963. That was the high point for the doomed Fermi I project.

On October 5, 1966, the plant "scrammed," or shut down suddenly and automatically. When engineers studied the accident, they quickly discovered that two fuel assemblies (out of more than one hundred) had melted. Much later they determined that two loose zirconium shields obstructed the flow of sodium. There was no radioactive contamination outside the reactor, but the fuel-melt was sobering, demonstrating that pushing a tricky and dangerous technology took untoward risks. As Mazuzan and Walker wrote, "A meltdown is the most critical problem any reactor can have. When the fuel melts, its behavior can become unpredictable."

Unscrambling the meltdown mystery and repairing the damage took four years. According to an analysis by the Union of Concerned Scientists, "It took months to remove the suspect fuel assemblies from the reactor core and ship them to Battelle Memorial Institute in Ohio for examination. It took longer to decontaminate the sodium loop and reactor components, drain the sodium loop sufficiently to probe for the cause of the meltdown, locate the offending parts, remove them, and put everything back together."

In May 1970, the plant was ready to restart. A sodium explosion halted the startup until July. The plant reached full power again in October 1970; during 1971 it operated fitfully, averaging only 3.4 percent of full power during the year. PRDC, foreseeing the end, refused to buy new fuel for the plant. In August 1972, the AEC denied an extension of the operating license. The plant shut down for good in September 1972.

The AEC's attitude toward the problems at Fermi I was, to put it gently, uncomprehending. The UCS analysis critiqued the agency for acting like a "hall monitor." It continues, "The agency passively reviewed the reports submitted to it by the Power Reactor Development Corporation and sent the occasional inspector to the site apparently more to satisfy geek needs like checking out the periscope developed to 'peer' through the sodium for the mysterious foreign objects than to audit recovery efforts. There was no discernible regulatory oversight by the AEC before this event or during recovery from it."

One of the most clear-sighted and practical pioneers of nuclear power was Adm. Hyman Rickover, the father of the nuclear Navy and the creator of the USS *Nautilus*, the Navy's first nuclear submarine. He had early hands-on experience with liquid sodium–cooled breeder reactors in the USS *Seawolf*, the Navy's second nuclear boat. By 1956, Rickover had already decided to scrap the sodium breeder power plant and replace it with a light-water reactor like the power plant in the *Nautilus*. Sodium-cooled systems, Rickover said, were "expensive to build, complex to operate, susceptible to prolonged shutdown as a result of even minor malfunctions, and difficult and time-consuming to repair."

Nonetheless, the Atomic Energy Commission, still deluded by the myth of uranium scarcity, was betting heavily, with taxpayer money, on breeder reactors. Any problems encountered along the way were, in the AEC's views, just technical, engineering problems that could be easily overcome. Thus came about the Götterdämmerung for breeder reactors in America: the Clinch River Breeder Reactor Project.

The skeptical views of its advisory committee and the problems at Arco and Fermi I in no way diverted the AEC away from its breeder enthusiasm. Inside the AEC staff, Milt Shaw, the long-time director of reactor technology, was the leading advocate for breeder reactors and what many had begun calling a "plutonium economy." In 1968, in an assumption-laden economic forecast of the need for breeders, Shaw wrote, "We have the potential to satisfy energy needs of mankind for a very long time by the process of breeding." During the late 1960s, Shaw curtailed safety research and development funds for conventional light-water reactors in order to boost spending on breeders. Shaw assiduously worked with the congressional joint committee, where California Democratic Rep. Chet Holifield, a long-time member of the joint committee, took the vows in the church of the holy breeder. Holifield termed breeders "indispensable."

The commission also wrapped itself in the breeding blanket. Chairman Glenn Seaborg proclaimed breeders "a priority national goal" and "the most decisive single step that could be taken now toward assuring an essentially unlimited energy supply, free from problems of fuel resources and atmospheric contamination." Seaborg took that enthusiasm to the White House, where, in an April 1971 meeting, he persuaded President Richard Nixon to go all-out for breeders. Nixon was preparing to deliver an energy message to Congress later that spring.

On June 4, 1971, Nixon's message went to the Capitol. Asserting that "a sufficient supply of clean energy is essential if we are to sustain healthy economic growth and improve the quality of our national life." Nixon outlined a series of steps his government would take, starting with "a commitment to complete the successful demonstration of the liquid metal fast breeder reactor by 1980." A year earlier, Nixon had proposed creating a new, overarching agency called the Department of Natural Resources to swallow the Interior Department and major programs from a dozen other cabinet-level agencies, as well as independent agencies such as the AEC. In his "Special Message to the Congress on Energy Resources," Nixon proposed "a single structure within the Department of Natural Resources uniting all important energy resource development programs."

Creating the new department was probably more important to Nixon than the breeder reactor demonstration project, but he needed to back the breeder to win Holifield's backing for the reorganization. Not only was Holifield a key member of the Joint Committee on Atomic Energy, but he was chairman of the House Government Operations Committee, which would have to approve Nixon's reorganization plan. The Department of Natural Resources quickly sank into the mists of administrative history as another failed reorganization, while the breeder program continued for more than a decade, gobbling up government funds, before it, too, failed.

A 1972 AEC memorandum of understanding laid out the plans for the breeder. It was followed by a series of detailed contracts among the federal government and the Tennessee Valley Authority (always a patsy for anything atomic and free of state or federal economic regulation), Chicago's Commonwealth Edison Co., and the Breeder Reactor Corp., with Westinghouse as the builder of the reactor. The plans called for construction to begin in 1974 and power generation in 1981.

The site for the demonstration reactor,[102] planned to have an electric capacity of 350 MW, would be some twelve hundred acres of AEC land on the Clinch River near the Oak Ridge weapons laboratory. The $699 million Clinch River Breeder Reactor would be a sodium-cooled reactor, fueled

---

102 As a "demonstration reactor," the breeder wasn't expected to meet full commercial expectations, but to show to potential private-sector investors that risks to an investment were manageable. As a practical matter, the government intended that the reactor would show the private sector that they would be fools not to build their own breeders. In failing, the reactor showed that the fools were found in government.

with a mixture of uranium and plutonium, and breeding more plutonium than it used in fuel. The Associated Press reported, "Private utilities across the country have pledged $240 million toward the project. TVA is putting up $22 million, and Commonwealth Edison $11.4 million. The balance will be supplied by the atomic agency."

The project seemed uncontroversial at the time but soon was generating plenty of political heat. Plutonium was the first source of friction. It didn't take long for many to realize that the plutonium economy would result in the spread of nuclear bombs around the world. Plutonium was a far better tool for building A-bombs than uranium, requiring only a small amount of the element to produce the guts of a nuclear weapon. In 1974, India, which had refused to sign the Nuclear Non-Proliferation Treaty, using plutonium reprocessed with U.S. help through an Atoms for Peace project, exploded an atomic bomb. The path from the peaceful atom to weapons of mass destruction went through plutonium reprocessing. In October 1976, President Gerald Ford put commercial plutonium reprocessing on hold. In March 1977, newly-installed President Jimmy Carter made Ford's hold permanent and also moved to kill the Clinch River breeder.

Cost escalation—the bane of nuclear programs throughout history— was also undermining the Clinch River project. The AEC's 1972 cost estimate for the project was $699 million, with the United States paying $422 million, and the private utilities covering the rest. When Westinghouse and the AEC finished the reference design for the plant, the cost estimate had risen to $1.7 billion. The private sector contribution was capped under the contract with the AEC. The taxpayers would have to eat the excess. By the time Congress finally drove a political stake through the heart of the breeder in 1983, the cost estimate topped an eye-popping $4 billion.

The endgame for the breeder lasted nearly seven years. The Carter administration proposed killing the project in 1977, but regional politics and the prospects for lavish pork barrel spending, combined with the administration's inept and cantankerous relations with Congress, kept the project alive but not thriving throughout the Carter years.

Some warning signs for breeder boosters came when ideological conservatives began questioning the logic of the government performing research and development programs for large, well-financed private companies. A young, hard-line GOP congressman from Michigan, David Stockman, in a

1977 press release slammed the Clinch River project as "totally incompatible with our free market approach to energy policy" and "a large uneconomic subsidy" to the nuclear industry.

When Ronald Reagan made Carter a one-term president in 1980, and the Republicans captured the Senate, the breeder boys were ecstatic. Tennessee's Republican Sen. Howard Baker, the most ardent supporter of Clinch River in Congress, became Senate Majority Leader. In the House, where the Democrats still ruled, Tennessee Rep. Marilyn Lloyd Bouquard, whose district included Oak Ridge, became chair of the House Science Committee's energy subcommittee and was a senior member of the Public Works Committee. There were a few small clouds on the horizon. One was Stockman, the spending hawk. He had become Reagan's budget director.

The *New York Times* in March 1981, as Reagan's first budget message went to Congress, reported from Oak Ridge, "Soon, however, the deer and the cottontails will probably have to move over and make a little room, because this is the site of the much-debated and much-delayed Clinch River Breeder Reactor, which President Carter tried to kill but only wounded in his fight to halt proliferation of nuclear weapons. But the political upheaval that recently shook the ground four hundred miles to the northeast in Washington appears to mean that the breeder—which makes, or 'breeds' more plutonium fuel than it consumes while generating electric power—will finally be built."

In a great political irony, a solidly-Democratic Congress refused the plan of the Democratic president to shut down a costly and vastly over-budget spending project that also threatened to upset the international nuclear non-proliferation regime. Then, a Congress with a Republican Senate and a staunch conservative in the White House killed the breeder. Stockman's 1977 critique foreshadowed a key reason. Political conservatives abandoned Clinch River. In an influential 1982 paper—"The Clinch River Folly"—Henry Sokolski, a young analyst for the Heritage Foundation, a right-wing think tank that had strongly supported Reagan's election, wrote:

> For many conservatives, Clinch River represents a dilemma. They are, on the one hand, strongly supportive of nuclear energy, but they are also concerned about the burgeoning federal deficit. Their opposition to the Clinch River Breeder, therefore, is born more out of a

concern to limit federal spending than opposition to nuclear power. The stakes are high. If, as spokesmen for the nuclear industry contend, the death of Clinch River will lead inevitably to the death of nuclear power in the United States, conservatives would undoubtedly continue to support the project. But the validity of this argument has become steadily more doubtful as new uranium discoveries, a general slowdown in the construction of nuclear power plants, and increasingly large cost overruns bring the wisdom of supporting Clinch River into question.[103]

Facing a powerful right-left coalition seeking to kill Clinch River, Howard Baker mounted a furious campaign to save the project in his home state. But early in the morning of October 26, 1983, the Senate dealt the death blow to the project, voting 56–40 against an appropriation of $1.5 billion for the staggering breeder program, on top of some $1.7 billion that had already been spent, with nothing to show for the expenditure. Sen. Dale Bumpers, an Arkansas Democrat and member of the Senate Energy Committee, said, "We put some money down a rat-hole and decided not to spend any more. It was a very dangerous course. The technology is considered by some respected scientists as less than viable."

The path to endless energy through a plutonium economy proved to be a very expensive dead end for U.S. taxpayers. A conservative 1975 General Accounting Office estimate put AEC liquid metal fast breeder reactor spending for fiscal years 1948–1974 at $1.8 billion in 1975 dollars. Clinch River nearly doubled that figure.[104]

On December 15, 1983, the U.S. Nuclear Regulatory Commission, one of the successors of the AEC, revoked a preliminary construction license for the Clinch River Breeder Reactor it had issued the year before. "With this

---

103 Sokolski joined the staff of New Hampshire Republican Sen. Gordon Humphrey, a leader of the anti-Clinch forces in the GOP, then the staff of Arizona Republican Sen. Dan Quayle. When Quayle became President George H.W. Bush's vice president, Sokolski joined the Pentagon staff as an expert on nuclear non-proliferation.

104 These estimates are certainly too low. For example, a GAO report in 1976 put the cost of the AEC's Fast Flux Test Facility in Hanford, Washington, not a breeder reactor but a key component in its development, at $1.1 billion.

action," wrote a group of academics studying the history of breeder reactors around the world in 2010, "breeder reactor development in the United States essentially ended."[105]

Breeder enthusiasm was built on bad assumptions by policymakers and resultant bad policy. In a 1972 paper, promoting the notion that low-cost uranium would soon run out, justifying breeders at nearly any cost, AEC head of reactor development Milt Shaw highlighted the problem with the nation's heavy breathing over breeders: "It should be noted that analytic studies which extend fifty years into the future should be used primarily to indicate trends that may result from changes in parameters. The validity of the projections is directly dependent on the validity of the assumptions used in the study. The reader should keep this fact and the assumptions clearly in mind when reviewing the results and avoid a natural tendency to use such parameter studies that involve projections into the future as absolute forecasts."

---

105 Breeder reactors have proven accident prone and expensive in other countries where the enthusiasm actually brought about working machines. France developed two breeder demonstration projects, neither of which proved economically viable. Japan's breeder had a major sodium fire and has been closed for over a decade, with attempts to resurrect it colliding with the March 2011 earthquake and tsunami that destroyed much of the country's conventional nuclear program and undermined nuclear power throughout the world. Germany abandoned breeder technology in 1991. The site of the German breeder, at Kalkar, is now an amusement park.

Waste is a Terrible Thing to Mind

In Washington in 2011, this joke was circulating among the followers of nuclear energy policy:

Question: What do you call it when you dig a hole in the desert, dump $15 billion in it, and walk away?

Answer: Yucca Mountain Nevada.

We return to the high, dry western desert where so much of the nation's atomic folly took place and so much of it was buried. The masters of the early, heady days of the Manhattan Project paid little mind to the nasty stuff left after the processes were refined, the bombs built, and the reactors constructed. They left behind a noxious mess of toxic and radioactive liquids, pumped into thin-walled tanks and largely forgotten. Hardly anyone contemplated what to do about the coming stream of detritus from the anticipated flood of commercial nuclear power plants. Over sixty years, this waste became an insoluble problem, an obstacle to any future development of civilian nuclear generation and a looming financial burden to commercial use of nuclear energy

Some .nuclear engineers like to dismiss the waste controversy as merely a "political" issue, as if that somehow meant it isn't significant and not worth puzzling over. First, that's not entirely true. There are plenty of technical and engineering issues that stand in the way of resolving the problem of what to do with what's left when the atoms have been split or fused, the lights lit up, and the Leaf's batteries run down. More to the point, the somehow unworthy "political" issue has helped erect a so-far insurmountable hurdle to an atomic future. No political resolution has come into view.

Samuel Walker, the in-house historian of the U.S. Nuclear Regulatory Commission, titled his short, pointed history of nuclear waste policy *"The Road to Yucca Mountain."* That is partly fitting, in that the nuclear dead end terminates at a hole in the ground located in a mountain in that old familiar haunt, the government's Nevada Test Site. But "road" is a bit misleading. The course that ends up in a multi-billion-dollar hole in the ground in the wastelands of Nevada is more like a meandering stream, or a confused cow path, with twists and turns, and no clear indication of its ultimate destination, if there be one.

20. Out of Sight and Mind

"Sanitary engineer" is a title that sounds anachronistic in an age of high energy physics, artificial intelligence, genomic sequencing, and Twitter. Sometimes, the phrase is a comedic euphemism for a garbage collector. Today, the more common term to describe such men and women is "environmental engineer," which doesn't really capture the heart of this honorable discipline.

Sanitary engineers are crucial to our modern world. These are men and women who make sure the water we drink won't sicken or kill us, that our waste byproducts won't diminish or damage our lives, that pestilence won't devastate our cities or overwhelm our populace.

Growing out of the craft of plumbing that dates at least to Roman times, modern sanitary engineering as a formal profession has its roots in 19th Century England. In laws in 1848 and 1858, the British Parliament established the basis for scientific regulation of public health, calling for investigations when a local death rate exceeded 23 per 1000 population. In an attempt to control smallpox, the Privy Council ordered the first systematic inspection of localities, which looked at public health in 1,143 districts in England in 1864.

Two British "Nuisances Removal Acts" in 1855 and 1860 establish ways for localities to appoint committees to deal with waste disposal, crowding, and other conditions that could have an impact on the spread of diseases among the public.

In the U.S., plumbing inspections became a common feature in cities in the late 19th and early 20th Centuries. In 1906, Henry Davis, chief plumbing inspector for the city of Washington, D.C., met with 25 of his colleagues from around the country at the historic Willard Hotel in the nation's capital to form the American Society of Sanitary Engineering for

Plumbing and Sanitary Research. Davis told the men,[106] "You are here for the purpose of forming an association of inspectors of plumbing and sanitary engineers, to be the source from which rules and regulations could be developed for the advancement of sanitary science, in the interest of public health." That organization survives today as the American Society of Sanitary Engineers.

The towering figure in American sanitary engineering was a little-known scientist from Baltimore, Maryland, Abel Wolman. Born in 1892 of immigrant parents from a Polish ghetto, Wolman never abandoned his native city. He graduated from Baltimore City College in 1909, earned a BA from Johns Hopkins University in 1913 in pre-medical studies, and a BS in engineering from Hopkins in 1915, one of the first four students to earn an engineering degree from the famous university. Wolman then began working for the Maryland Department of Health, where he stayed until 1939, when he became a full-time professor and department chairman at the Hopkins Department of Sanitary Engineering, which he had founded in 1937.

Wolman's claim to lasting fame came in 1919, at the age of twenty-six, when he devised a method for standardizing the use of the deadly chemical chlorine to purify drinking water.[107] He successfully applied the new technology to Baltimore's water supply, making it a model for the rest of the nation. That began a long career as a consultant to municipalities and institutions around the world on public health and waste handling and disposal.

Wolman's work on chlorination, still the primary method for water purification around the world, was a revolutionary development in public health. Potable water has transformed the world. In 1997, *Life* magazine called water chlorination "probably the most significant public health advance of the millennium."[108]

In 1948, during the public relations glow resulting from the successes of the Manhattan Project, Wolman also became the first prominent scientist to warn of the dangers of the growing accumulation of nuclear waste

---

106 Of course, they were all men.

107 Wolman worked with Hopkins chemist Linn H. Enslow on chlorination

108 In 1987, the American Public Works Association created the Abel Wolman Award for the best new book published in the field of public works history.

from the nation's atomic bomb program. In raising a warning about the potential public health impacts of nuclear waste, Wolman found himself in conflict with the powerful nuclear establishment, including the iconic physicist J. Robert Oppenheimer, chairman of the Atomic Energy Commission's General Advisory Committee. While never gaining the public fame or notoriety of scientists such as Oppenheimer or Edward Teller, Wolman undoubtedly had a greater impact on the life of the average American—or the average citizen of the world, for that matter.

The pioneers of the Manhattan Project paid little attention to what was left behind when they stripped U-235 from U-238 or made plutonium from uranium—understandable at the time, although shortsighted in retrospect. They faced and felt great urgency. When they thought about waste at all, they thought it was a simple technical problem they could deal with in time. The atomic energy establishment came up with a revealing term to describe the work of understanding the impacts of radioactivity on humans and the environment. This discipline quickly became known as "health physics." One observer suggested that the term was concocted as a result of the cult of secrecy during the early days of the bomb program: "'Radiation protection' might arouse unwelcome interest; 'health physics' conveyed nothing."

The major source of waste from the Manhattan Project was a wicked liquid brew of chemicals and radioactive isotopes resulting from the plutonium reprocessing operation at the Hanford Engineering Works on the Columbia River in Washington. The bomb dropped on Nagasaki in 1945 was fueled with plutonium. Operations at Hanford consisted of: reactors that bombarded aluminum-clad uranium-238 fuel pellets with neutrons, creating, among other byproducts, the necessary Pu-239; and the plants that stripped the plutonium from the irradiated reactor fuel.

To get the irradiated fuel pellets from the Hanford reactor to yield its explosive plutonium, workers first moved the hot fuel to a pool of water to cool down and age sufficiently to let radioactive iodine-131 decay to less dangerous levels.[109] Then the irradiated slugs were moved to a reprocessing building, soaked in a chemical bath to remove the aluminum cladding

---

109 I-131 has a half-life of eight days, meaning it loses half of its radioactivity in just over a week.

around the fuel, chopped into bits, and dissolved in a bismuth phosphate solution. A centrifuge removed the plutonium, producing a plutonium nitrate paste that Hanford then sent to Los Alamos for use in bombs.

The plutonium extraction took place in large, long concrete industrial buildings known as "canyons." The most famous, the T plant, where the Hanford technicians removed the plutonium for the Nagasaki bomb, was eight hundred feet long, sixty-five feet wide, and eighty feet high. The Hanford workers named it Queen Mary, for its general resemblance to an ocean liner.

What was left after the Hanford technicians removed the plutonium was a thermally and atomically hot, nasty mix of toxic and radioactive chemicals, capable of boiling by itself, highly corrosive, and not reusable. Not knowing how to convert this witches' brew into something useful or even stable, the plant's engineering contractors—the chemical company DuPont designed, built, and operated the plant—provided a tank farm for waste storage. They initially built sixty-six underground tanks. The bottoms and sides of the tanks were carbon steel wrapped in reinforced concrete. The tanks were topped with concrete lids. The first tank farm had a capacity of about twenty-five million gallons. Another eighty-three single-shell tanks went into the ground after the war, up until 1964. From 1968 to 1986, Hanford engineers built another twenty-eight tanks. The final generation of Hanford tanks had double shells, with a second steel container wrapped around the first carbon steel and concrete shell.

But even with hundreds of millions of gallons of storage capacity, the Hanford tank farms were far too small to handle all the liquid waste the plant generated, particularly after 1952 when a new, high-volume Redox plutonium extraction process went into service. By early 1953, Hanford was generating some nine million gallons of liquid waste per day. As the Hanford official website[110] delicately explains, "The volume of chemical wastes generated through the plutonium production mission far exceeded the capacity of the tanks. Some of the liquid waste did end up being put into holding facilities and some was poured into open trenches. Some of the wastes that were put into the tanks didn't stay there, as the heat generated by the waste and the composition of the waste caused an estimated

---

110 http://www.hanford.gov/page.cfm/TankFarms

sixty-seven of these tanks to leak some of their contents into the ground. Some of this liquid waste migrated through the ground and has reached the groundwater."

The men in charge of the mammoth enterprise appeared unconcerned about this approach to waste. Herbert Parker, head of health and safety programs at Hanford, argued in the late 1940s that "present disposal procedures may be continued...with the assurance of safety for a period of perhaps fifty years." Parker, a British radiologist, believed that his mission would be accomplished if the radioactive and chemical wastes simply remained isolated on the Hanford federal reservation. "What the water does at Hanford," he explained, "is irrelevant if the toxic matter is retained." As University of Montana historian Ian Stacy put it in a 2010 article in the journal *Environmental History*, "Apparently, then, the only way to preserve the regional ecosystem was to destroy the Hanford reservation."

For decades, officials at Hanford, the Atomic Energy Commission, the Energy Research and Development Administration, and the Department of Energy in Washington insisted that the tanks would not leak. When it was shown that they were leaking, the officials said the structure of the soil and the fact that the tanks were well above the ground water meant the contamination could never move offsite. When it was shown that Hanford was polluting the Columbia River, the officials effectively said, "Well, never mind. We'll figure out a way to fix it, and get back to you."

The atomic enterprise's chief concern about waste produced from its operations was the laudable aim of keeping worker exposures to minimal levels.[111] The weapons engineers were also concerned about economic damage to livestock and, in the case of Hanford, to economically important salmon stocks in the Columbia. In a wonderfully understated comment, a 1997 Department of Energy look-back at nuclear waste management said, "The Manhattan Project officials did understand, at least theoretically, the environmental, health, and safety implications of nuclear technology."

When the AEC began its operations, taking over from the Army, the agency created a Safety and Industrial Health Advisory Board, composed

---

111 The issue of what is an acceptable dose of radiation to workers, and to the public, has always been evolving. What was considered acceptable in 1945 was far greater than what most experts would consider tolerable today.

of outside experts, with a mission of examining the Manhattan Project's fire, construction, chemical, electrical, and radiation hazard programs. Abel Wolman in the late 1930s had done some consulting work for the Tennessee Valley Authority, under the direction of TVA chairman David Lilienthal. When President Truman named Lilienthal chairman of the Atomic Energy Commission in 1947, it was natural that Wolman, who was also the chairman of the National Academy of Sciences' Committee on Sanitary Engineering, would be on the AEC advisory board.

Wolman probably was behind the creation of the AEC board. The academy originally asked him to visit Lilienthal to express its concerns about the agency's management of nuclear waste. As chairman of the relevant NAS committee, Wolman was almost certainly the source of the academy's interest in the topic. He advised Lilienthal that the AEC needed to foresee worker and public health problems that might "arise in the continued development of nuclear fission studies and production programs." Lilienthal responded that he thought that the academy's "worry was unwarranted." But the AEC chairman created the advisory board and, entirely predictably, named Wolman to serve on the board, which was headed by Sidney Williams, assistant to the president of the National Safety Council.[112]

The indefatigable Wolman brought his full energy to the AEC work. He boldly went into the field, visiting the AEC's labs and production plants, including Hanford. In 1947, working with another sanitary engineer, his friend Arthur Gorman, Wolman produced a draft report on the AEC facilities for the advisory board. Wolman and Gorman got to the heart of the AEC's waste problems, noting that "in the haste to produce atomic bombs during the war certain risks may have been taken…with the understanding that subsequently more effective control measures would ameliorate those risks." The limits for exposure to radiation and toxic substances, they said, were set internally, without consulting with "public health officers normally concerned with and responsible for such problems in civilian life." Wolman and Gorman feared that water supplies at Hanford and the Clinton lab in Oak Ridge had been contaminated, and that air pollution

---

112 The NSC was a private-sector group, formed in 1913 as the National Council for Industrial Safety. In 1953, President Eisenhower and Congress recognized the group with a government charter.

was a significant problem. "We cannot recall a single stack in any of the areas of such height or design which would meet modern requirements of industrial or laboratory operations," they concluded.[113]

The safety advisory board's final report echoed the analysis of Wolman and Gorman, although in somewhat more diplomatic language. "The Atomic Energy Commission inherited from the Manhattan Project an excellent safety program and record," said the final 1948 report. But then the board slid in its analytic knife: "There are recent indications that these are deteriorating."

The board delivered its final report to Lilienthal and the AEC in April 1948. A 1997 Department of Energy look at the history of radioactive waste management, summarizing the Wolman panel, characterized the AEC's waste management program as "negligent." Continuation of AEC waste disposal practices, said the Wolman board, presented "the gravest of problems" in the long run. The board concentrated on gaps in the AEC's understanding: how its liquid wastes would migrate toward groundwater, streams, and rivers; how the AEC production facilities were venting gaseous radioactive and chemical wastes; and further studies of the impact of waste disposal on the surrounding environment.

The advisory board also recommended that the AEC reorganize its activities to increase its attention to waste. Under Manhattan Project protocols, waste disposal was entirely a local matter. The managers of the labs and atomic factories were responsible not only for what they produced but how they disposed of the remaining noxious material. The 1997 report on radioactive waste management noted: "The Hanford health physics capability was set up independent of the Met Lab[114] in Chicago. Los Alamos also established and conducted its own health physics program. High security hampered cooperation and data exchange among the laboratories' health physicists." The safety board said this undervalued waste manage-

---

113 In 1949, Gorman and Wolman published an important paper, "Some Public Health Problems in Nuclear Fission Operations," in The Nation's Health, the journal of the American Public Health Association.

114 The Metallurgical Laboratory in Chicago, where Fermi built his radioactive pile, was the predecessor to the Argonne National Laboratory.

ment, producing a concentration on radiation hazards to workers, with not enough attention on release of toxic wastes into the environment.

Understanding the history of local autonomy and authority that characterized the atomic endeavor, the NAS advisory board tiptoed carefully around the issue of management structure. The board recommended a "gradual" approach to centralization, slowly concentrating authority and oversight for environmental safety and health into a headquarters function centered in Washington.

The Wolman board's report had mixed impact. For the most part, the physicists in charge of the atomic policy rejected the board's notion that there was something rotten in the AEC's waste management approach. Wolman quickly crossed policy swords with the legendary Oppenheimer, then at the height of his bureaucratic and public relations powers. At the April 1948 meeting where the board discussed its findings, Oppenheimer dismissed waste disposal as "unimportant." Historian Samuel Walker said Oppenheimer's attitude was "a prevalent judgment among physicists and other scientists who held influential posts with the AEC."[115]

The AEC bureaucracy also rejected the idea of centralizing waste management. A group representing the AEC's labs met in May 1948 and voted in favor of continuing the policy of leaving authority for waste disposal in the hands of local managers. That view, however, was not unanimous. Walter Zinn, the new head of the Argonne National Laboratory outside Chicago and the leading light of the agency's breeder reactor program, disagreed. He told his staff in 1948 that radioactive waste disposal was one of the greatest problems facing the lab and should be on the agenda of the AEC in Washington.

Wolman and the safety board did persuade Lilienthal to take some small steps toward centralizing the commission's environmental safety and health program. He agreed to what the report proposed as a first step: appointing a sanitary engineer to the AEC's headquarters staff. Lilienthal told Wolman, "Since you pushed it, you find me the man."

---

115 Walker recounts a later dinner party at Oppenheimer's home, where an un-cowed Wolman told the physicist that "your ideas as to how we shall manage this 'unimportant' problem are characterized almost completely by a total ignorance of the nature of disposal."

Wolman had no trouble finding the man. He immediately nominated his close friend and colleague Arthur Gorman for the job. Gorman was running the city water department for Chicago but agreed to take the job at the AEC. Gorman then, in 1949, hired Joseph Lieberman, who had earned his PhD in sanitary engineering under Wolman at Johns Hopkins in Baltimore, to assist him. As NRC historian Walker summarizes, "For many years, Gorman and Lieberman were the entire sanitary engineering staff of the AEC, and in that capacity they sought, in the face of both technological and political hurdles, satisfactory ways to deal with growing quantities of radioactive waste."

21. Holey Kansas

"Backbencher" is a political description, derived from British experience, where junior and unimportant members of Parliament sat on benches at the back of the legislative chamber. In the U.S., the term has come to describe members of the U.S. House of Representatives who do not control or contribute much to the high-profile, national work of the institution. Rather, they are the legislators who focus on hometown issues, bring home the federal bacon, and depart without having made much of a mark on the nation.

Backbencher Joe Skubitz, a middle-of-the road Republican from the middle of Kansas, largely stuck to his political knitting and didn't make much noise. Skubitz served the fifth district of Kansas from 1963 to 1978. A member of the minority party during a period of long Democratic hegemony, he never chaired a House committee. He had a dependably Republican voting record, an average score for attendance in the House, and left little mark on the institution or the United States—except when it came to nuclear waste.

Skubitz wouldn't have been able to make the kind of ruckus and arouse the kind of interest it took to derail the nation's first serious attempt to deal with nuclear waste had it not been for the hubris and disregard for local citizens that long characterized the U.S. nuclear establishment, in Congress and the Atomic Energy Commission. The course of events that led to the collision between the backbencher and the nation's atomic energy juggernaut began in 1954. Responding to President Eisenhower's Atoms for Peace speech, Congress passed the second Atomic Energy Act. The key thrust of U.S. nuclear policy shifted perceptibly from bombs and weapons to producing electricity.

The Atomic Energy Commission recognized that development of a robust civilian power industry, however it occurred, would mean increased volumes of radioactive waste. The commission had already come under criticism for its handling of war wastes at Hanford, Oak Ridge, and its other weapons labs and factories. Now, the commission also had to consider wastes from power plants, although it saw this problem as one of degree, not nature.

Despite assurances from the field offices and laboratories that the AEC's management of high-level nuclear waste was prudent, conservative, and not a threat to anyone anytime soon, doubts stirred in the mid-1950s, intensified in the 1960s, and led to an embarrassing fiasco in Kansas in the 1970s. The affair demonstrated the continuing inability of the AEC and the nation to come to grips with the noxious residue of the friendly atom. The events also demonstrated the increasingly obvious fallibility of the nuclear priesthood.

At the AEC, the prospects of dealing with highly-radioactive used fuel from power plants didn't represent anything new from a technical standpoint. The commission and its staff assumed that the spent nuclear fuel would be treated as they handled the used fuel from weapons reactors. The fuel would be chopped up, dissolved in chemicals, and the resulting liquid wastes stored in underground tank farms until it could be turned into some form, not yet settled, of solid waste. The waste problem created by civilian plants would be a matter of quantity.

In 1955, Glenn Seaborg said that disposing of the "tremendous quantity" of waste "may well be limiting factor" for the growth of civilian atomic energy. But Seaborg expressed the characteristic optimism of the time. "These problems will be solved, however," he said, "and a nuclear energy will probably be developed in the future because of the advantages of this form of energy."

The next year, the National Academy of Sciences panel chaired by Wolman called radioactive waste "an unparalleled problem." While the NAS report concluded that various disposal technologies might work, much more work would be required "before any of them is at the point of economic operating reality." A year later, an academy report found that putting highly radioactive wastes in salt formations was the most promising course for disposing of nuclear waste.

The AEC asked Wolman's NAS committee to recommend the best way to dispose of the volume of high-level waste it had already accumulated at its weapons sites as well as what was expected to pile up at the civilian power plants. In largely hortatory language, the Committee on Waste Disposal advised that "the hazard related to radio-waste is so great that no element of doubt should be allowed to exist regarding safety." The committee fingered salt deposits as the best way to hide the high-level wastes.

Salt, said the NAS committee, had many advantages. Large salt deposits tend to be dry, and water is the great enemy of nuclear waste disposal, as it provides the path for the radioactive elements to escape into the environment. Cracks and fissures in salt beds tend to take care of themselves; salt is plastic and moves on its own to close up gaps. The largest salt beds in the U.S. are located where there are few earthquakes and little volcanic activity to disturb the stored wastes over thousands of years. They generally are found in level formations, making underground excavation easier.

One area with salt formations was particularly attractive, because it was located in the center of the country, reasonably equidistant from potential reactor sites across the nation. That place was Kansas.[116]

During the 1960s, the AEC decided that the best way to deal with nuclear wastes, which it assumed would all be liquid products from reprocessing plants, was to dry out and solidify the material before it was put into steel containers and buried somewhere. The idea was to reduce the volume of the wastes by up to 90 percent. But daunting problems remained. The solidified waste would remain dangerous far longer than the projected life of the steel cans the AEC was proposing to use to hold it. The agency didn't know how the solidified, compacted waste in its containers would behave when the containers were placed close together underground. Would the

---

116 A small notice in the December 4, 1891 edition of the *New York Times* provides a sense of the size of the salt beds in central Kansas. "Chicago, Dec. 3.—A meeting of rock salt miners was held here yesterday for the purpose of taking preliminary steps for forming a combination. The following firms were represented: Roy Rock Salt Company, Kansas City; Crystal Rock Salt Mining Company, Chicago; Lyons, Kan; Midland Rock Salt Company, Cleveland; Kingman Rock Salt Company, Kingman, Kan. The mines worked by all of these companies are in Kansas, and the aggregate capital of the firms is $2,000,000. The output of the mines is 7,000 tons a day. These firms control all the rock salt mines in Kansas, and they propose to form a pool, with a working capital of $1,000,000."

heat remaining in the waste accumulate and damage its surroundings? How would water interact with the waste packages?

Waste researchers at Oak Ridge decided to test out some of these unknowns with simulated wastes in two abandoned salt mines in Kansas in 1960. By 1965, the commission felt the Oak Ridge research in Kansas had "shown strong promise" and the AEC was moving toward the Kansas salt beds as its final solution to the waste problem.

The AEC's focus on salt deposits in Kansas was not entirely a technical matter. The commission was also engaged in political skirmishing with the NAS and Abel Wolman's waste committee. At Wolman's suggestion, in 1964 NAS president Frederick Seitz, a distinguished physicist who had studied under Eugene Wigner and ran the atomic training program at Oak Ridge from 1946 to 1947, appointed an ad hoc group, led by Wolman, to look at the "state of the art of disposal of radioactive wastes." In mid-1965, the review chided the commission for failing to take a systematic approach to nuclear wastes and dealing with the problem piecemeal and day-to-day.

Seitz sent the ad hoc committee's comments to Glenn Seaborg, who parroted the atomic party line in response. The AEC was working on the problem, Seaborg said, and was considering "a long-range comprehensive waste management plan." The subtext of Seaborg's response was characteristic of the AEC: leave us alone, and we'll take care of it. Milt Shaw, one of the AEC's top permanent staff and head of the reactor development division, tried but failed to get the academy's nuclear waste panel abolished.

The attention from the academy, although out of public sight, pushed the AEC to move faster on waste burial. The agency increased the size and pace of its activities in Kansas. Between 1965 and 1968, the commission ran a series of experiments using spent fuel elements from the reactor testing site in Idaho to see how solid waste and salt would work together. The experiment, known as Project Salt Vault, took place in an abandoned mine in Lyons owned by the Carey Salt Company.

Project Salt Vault, in the view of the researchers at Oak Ridge, was a success. The canisters of radioactive fuel from Idaho were loaded into casks, trucked to Kansas, lowered into steel-lined shafts over a thousand feet underground, and replaced every six months over a nineteen-month period. In early 1970, the AEC researchers reported that "most of the major technical problems pertinent to the disposal of highly radioactive waste in salt have been resolved." The AEC determined to move forward with a full-scale waste dump[117] in the salt beds near Lyons. The AEC figured it might as well use the abandoned Carey mine, where it had accumulated much data during Project Salt Vault.

Characteristically, the AEC officials involved in the project in Kansas had not taken care to ensure that the local public supported the project. The AEC staff reported to the commission that they had consulted with "principal officials" and those discussions appeared to "indicate support" for the waste site. They were fooling themselves and fooling the commission.

While many locals in Lyons looked on the project with lust for government money and jobs, much of the rest of the state was restive. Nuclear Regulatory Commission historian Samuel Walker summarized, "The prevailing attitude in Kansas as the AEC took preliminary action on the construction of a waste repository in spring 1970 was ambivalence." One of the state's leading newspapers, the *Topeka State Journal*, wrote that the state should consider the plan for Lyons "with more than the proverbial grain of salt." The state's popular Democratic governor Robert Docking was conflicted by the project.

Seeking a third term in the predominantly-Republican state, Gov. Docking asked William Hambleton, the state geologist and head of the Kansas Geological Survey, to brief him on the radioactive waste dump. Hambleton warned Docking that the project would be technically complex and politically controversial and that the AEC would attempt to bulldoze the opposition and get a quick authorization from Congress to move ahead with the project.

His assessment was right on target. On June 17, 1970, in Topeka, the AEC's waste project manager, lawyer John Erlewine, announced the selection of the Carey Salt abandoned mine in Lyons as the site for the nation's underground radioactive waste dump. According to an account by United

---

117 The nuclear establishment cringes at the term "dump," preferring "repository" as better reflecting the careful engineering and technology that will go into a final resting place for retired fuel. Former Nevada Republican Sen. Chic Hecht, known for his malapropisms, in the early 1980s referred to the Yucca Mountain project in his home state as a "nuclear suppository." I use the terms dump and repository interchangeably.

Press International, Erlewine's announcement came "following a meeting with the Kansas Nuclear Energy Council. The area was named the Kansas Nuclear Energy Park." Erlewine said the AEC would ask Congress for $25 million in fiscal year 1972 for the project.

The next month, Erlewine briefed the Wolman committee at a National Academy of Sciences meeting in Oak Ridge. He repeated what he had said in Kansas. The AEC had settled on Lyons for the repository and would ask Congress for $25 million in 1972 funds. He wanted to be able to tell the Joint Committee on Atomic Energy that the NAS supported the choice of Kansas and the salt mine at Lyons for the project.

Enter Joe Skubitz. While little known outside his home territory centered in Pittsburg, Kansas, Skubitz understood how Washington worked and knew how to gum up the levers of power.

Son of Slovenian immigrants—his father was a coal miner and a small merchant—Skubitz was a bright young man who taught at a Pittsburg's Mineral High School while earning a degree at Pittsburg State. In 1939, the 33-year-old Skubitz moved to Washington to work for Kansas Republican Senator Clyde Reed. While working as a Senate staffer, Skubitz, like many other congressional aides over the years, earned a law degree at night at George Washington University. He then went to work as a staffer for Republican Senator Andrew Schoeppel, who was skeptical of the AEC's 1960 experiments in Kansas salt beds.

Schoeppel died in 1962 and Skubitz decided to run for Congress as a long-shot candidate. Skubitz demonstrated his political chops in the 1962 election cycle. He first defeated Republican Walter McVey, the incumbent in the third Kansas congressional district, for the GOP nomination in the fifth district. Redistricting following the 1960 census put McVey and another, more popular and senior Republican incumbent, in the same district so McVey sought the GOP nomination in the fifth district. Skubitz knocked off McVey by fewer than 200 votes out of 30,000 cast in the August 1962 primary. Skubitz then defeated the three-term incumbent Democrat James Floyd Breeding in the general election. He was never seriously challenged again and retired at the end of 1978.

A savvy political pro, Skubitz was also a conventional Republican in his skepticism about the federal government and preference for state solutions to public policy problems. He took a position on the House Interior Committee, where he largely defended business interests and state authority while promoting national parks.

In the fall of 1972, for example, Skubitz clashed with conservationists and the Nixon administration's Interior Department leadership during a conference celebrating the one hundredth anniversary of Yellowstone National Park. A study commission by the National Park Service and written by the Washington environmental group Conservation International called for major restrictions on use of the national parks in order to preserve their natural heritage. Nixon's Interior secretary Rogers C.B. Morton, a former Maryland Republican congressman, supported curbing car traffic in the parks and moving commercial facilities outside the park boundaries.

Skubitz, joined by other Republican congressmen, denounced the report at the Jackson Hole, Wyoming, birthday celebration where the report was unveiled. "Parks are for people," Skubitz said. "All the people. They must never become totally commercialized. Neither should they be locked up under the guise that they can only be preserved by so doing."

Skubitz's skepticism toward the federal government came to the fore in his reaction to the plans to dump the nation's nuclear waste in the Lyons salt mine—despite the fact that the abandoned mine was more than one hundred miles from his home district. When he heard about the Topeka announcement by the AEC's Erlewine, Skubitz wrote Gov. Docking about his doubts and fears. "We are being asked to assume unknown risks to make Kansas a nuclear dumping ground for all the rest of the nation," Skubitz told Docking. His strategy was clear. He wanted to line up a wall of bipartisan opposition from Kansas to persuade Congress to slow down the rush to bury waste in Lyons.

The AEC unintentionally helped Skubitz make his case by producing a wholly inadequate environmental impact statement for the project in early 1971. The National Environmental Policy Act went into effect in 1970, requiring environmental analysis for "major federal actions." The AEC and the Joint Committee on Atomic Energy consistently attempted to narrow the application of the law to AEC activities, which the commission argued

211

were largely exempt from the reach of NEPA.[118] The AEC staff viewed environmental impact statements largely as a nuisance and a bureaucratic requirement that could be brushed aside.

The AEC's impact statement for Lyons predictably found "no significant impact"[119] in the construction or operation of the Lyons waste dump. Kansas state geologist Hambleton dismissed the EIS as "general, meaningless, and a public relations effort." Docking denounced it for avoiding "major problems" including transportation, geology, and monitoring.

Skubitz joined the chorus, sending long, detailed letters to Docking, AEC chairman Seaborg, and the Kansas legislature outlining his many objections to the project and to the process that the AEC was using to win approval of the Lyons waste dump. The local chapter of the Sierra Club released Skubitz's letter to Docking to the press, along with Hambleton's objections.

Based on its environmental analysis, the AEC asked the congressional joint committee for an immediate $3.5 million to begin site work at Lyons in fiscal 1972, along with a multi-year authorization of $27.5 million. The committee began two days of hearings on the AEC request on March 16, 1971, where Skubitz, not a member of the committee, carpet-bombed the AEC request with detailed objections. In some forty pages of testimony, the Kansas Republican directly confronted the AEC's attitude toward its skeptics, and specifically Chairman Seaborg. "I hesitate to differ with an eminent Nobel Prize winner," Skubitz said, "but I must in good conscience point out that it is this attitude of 'leave it to us; we're great scientists' that most affronts a layman. The AEC acts as if your concern and mine should be limited to acknowledging their superior intellect and following their dicta."

The objections of Skubitz, coupled with those of Docking and other prominent Kansas politicians, and a desire by Kansas Republican Senators

---

118 It took a major defeat for the AEC in 1971 at the U.S. Court of Appeals for the District of Columbia Circuit in the case of Maryland's Calvert Cliffs nuclear plant to clarify that NEPA fully applies to federal nuclear facilities.

119 A "finding of no significant impact" is known in the field of environmental law as a "Fonsi," pronounced like the name of the Henry Winkler character (Arthur Herbert Fonzarelli) in the 1970s TV sitcom Happy Days.

Robert Dole and James Pearson to craft a compromise, slowed down the AEC's express train to Lyons. The JCAE agreed with a plan worked out by Skubitz and Dole to establish an "advisory committee" including Kansas officials to oversee the work, and that the AEC could only lease, not own, the site.

As Congress was enacting this compromise, the AEC waste express derailed. In May 1971, the president of the American Salt Corp. of Kansas City, Missouri, one of the operators in the Lyons salt bed, wrote to Oak Ridge that its plans for the waste dump failed to account for many oil and gas wells that had been drilled in the region over the years. The residue of the drilling could provide a path for corrosive water to infiltrate the waste site. The salt company official, Otto Rueschoff, noted that drilling by his company in April accidentally released "several hundred gallons of brine and muck." In July, he told the Kansas Geological Survey, the state government body that Hambleton ran, that in 1965 the company had pumped 175,000 gallons into an injection well at Lyons as part of its mining process. The company was using a technology from the oil and gas industry known as hydraulic fracturing to increase production from the solution mining operation. The workers forced the water down a borehole under great pressure designed to break apart the salt and make it dissolve easier. Instead, the water simply vanished and the company experts were never able to determine where it went.

Pondering American Salt's revelations, Hambleton commented that "the Lyons site is a bit like a piece of Swiss cheese, and the possibility for entrance and circulation of fluids is great." Lyons was effectively killed, or, as Skubitz announced in September 1971, "dead as a dodo." While the AEC refused to admit that the dump had turned into a corpse, by early 1972 the commission announced that it would shift its waste focus to above-ground storage.

Joe Skubitz's 15 minutes of fame were over. He returned to his comfortable role as a reliable Republican backbencher, where he served for another six years.

The damage to the AEC was significant. Samuel Walker wrote, "The AEC paid a heavy price for its errors. The Lyons debacle received national attention that diminished confidence in the agency and made its search for a solution to the waste problem immeasurably more difficult."

M. King Hubbert, a prominent geologist and supporter of nuclear power who served on the academy's waste panel for many years and chaired it for a period, later told journalist Luther J. Carter, "There has never been an agency any more ruthless than the AEC. They were a law unto themselves. They had an entirely collusive Joint Committee on Atomic Energy, and there were simply no constraints." But they hadn't reckoned on Joe Skubitz.

22. The Jimmy, Ron, and Mo show

Its reputation in tatters from the Lyons fiasco, the Atomic Energy Commission responded with classic bureaucratic behavior, described best by academic and satirist James H. Boren: "When in doubt, mumble; when in trouble, delegate; when in charge, ponder."[120]

The commission mounted a literal waste holding action, focused on what new commission waste czar Frank Pittman viewed as a prudent approach to the unsolved and increasingly politicized problem. Pittman advised the commission that, for the short term, the best course of action was to keep the wastes safe in above-ground storage, where technicians could monitor it, and where it could be retrieved once a long-term solution came to pass. As to what the long-term solution would be, Pittman basically had no clue. He suggested that the AEC look at a variety of disposal options to be tested before making a final choice of the form of geologic disposal.

Pittman's approach appealed to the pragmatic new AEC chairman, James Schlesinger, the Nixon administration's replacement of Glenn Seaborg, who had served for a decade at the head of the agency. Schlesinger served at the AEC for less than two years, although that was far from his last encounter with nuclear waste and nuclear power issues.

---

120 James H. Boren, When in Doubt, Mumble: A Bureaucrat's Handbook, Van Nostrand Reinhold, 1972. Jim Boren (1925-2010), author, humorist, and sculptor, had been an administrative assistant to Democratic Texas Senator Ralph Yarborough and was a cousin of former Oklahoma Democratic Senator David Boren, now the president of Oklahoma University. Jim Boren was a professor of political science at Northeastern State University in Tahlequah, Texas. Among his works was How to be a Sincere Phoney, A Handbook for Politicians and Bureaucrats. He once ran for president under the slogan, "I have what it takes to take what you have."

As the AEC temporized on radioactive waste, the political ground was shifting under the AEC's foundation. Long before the commission was able to come up with a new strategy for disposal of its nuclear detritus, the commission itself ceased to exist. At the same time, the commission's assumption that civilian nuclear wastes would be reprocessed for their plutonium to be used in a mixture with slightly-enriched uranium in future civilian reactors also turned to sand.[121]

The golden age of civilian nuclear power that the AEC and the Joint Committee on Atomic Energy had long anticipated had only a very short run. By 1974, the long pipeline of nuclear plants began emptying. Plants on order and under construction began falling by the wayside, derailed by a slow economy, rising interest rates, political opposition, poor construction, and declining performance of existing plants.[122] The last order for a new plant came in 1978, and that was illusory.[123]

The blossoming anti-nuclear movement of the mid-1970s challenged the fundamentals of atomic energy in the United States, including the engine of nuclear power, the AEC. Critics charged that it made no sense for the government to combine promotion of the technology and its regulation in the same agency. The result, the critics correctly observed, was that regulation became the stepchild to the nuclear boomers in the AEC. Even many in the industry acknowledged that in the AEC, the regulators were second-class citizens. In 1974, Congress acknowledged that it was time to formally

---

121 Reprocessing, long a goal of the nuclear establishment, is not a solution to the nuclear waste problem. At best, it makes a marginal contribution by reducing the volume of the material to be stored in the waste dump, assuming that the liquid mess from reprocessing can be solidified. It can also reduce the astonishingly long period of time necessary for isolating the waste. But reprocessing also adds greatly to the cost of storage, as well as providing another path for undesirables to obtain plutonium for clandestine nuclear weapons programs.

122 The devastating March 22, 1975, fire at the TVA's Browns Ferry reactor, at the time the largest nuclear power plant in the world, did much to undermine the myth of infallibility of the nuclear priesthood in Washington and the safety claims of the industry.

123 Commonwealth Edison in Chicago, the nation's largest private-sector nuclear utility, ordered two companion units in 1978 but never applied for a federal license to build the plants or took any action to indicate the orders were serious. The company quietly wiped the orders off its books in the mid-1980s.

bury the offspring of the Manhattan Project (although the Manhattan Project culture survives to this day). President Gerald Ford on October 11, 1974, signed the Energy Reorganization Act, abolishing the AEC. In its place, Congress created two separate federal agencies. The Energy Research and Development Administration (which became the foundation for the Department of Energy in the Carter administration) took over the promotional and research and development aspects of the AEC, along with the AEC's nuclear weapons program. A five-member U.S. Nuclear Regulatory Commission was charged with the task of being an independent and unbiased regulator of civilian nuclear power and use of the atom. Jurisdiction over nuclear waste, both military and civilian, was divided and unclear.

Equally murky was the relationship between the two nuclear agencies and the U.S. Environmental Protection Agency, created in the Nixon administration just four years earlier.

During the 1976 presidential campaign, energy in general and nuclear reprocessing specifically became an issue, with the Democratic candidate, Georgia's Jimmy Carter,[124] proclaiming that he would end the use of the fuel processing technology. Reprocessing, he told a United Nations energy conference, represented too easy a path for terrorists and rogue nations to obtain nuclear weapons. He pledged to put U.S. reprocessing on hold if he were elected.

President Ford in the summer of 1976 ordered a full-fledged review of U.S. nuclear policy, headed by Bob Fri, deputy administrator at ERDA, and coordinated by Glenn Schleede, a former AEC staffer by then on the White House staff. As a result of that work, on October 28, 1976, days before the election, Ford announced an indefinite moratorium on reprocessing spent nuclear fuel.

When Carter won the election, reprocessing in the United States died, and the focus of the still-inchoate nuclear waste policy changed from solidifying liquid wastes to dealing with the hot, radioactive spent fuel rods from civilian nuclear power plants. For the utilities operating nuclear plants, this

---

124 Carter is a nuclear engineering graduate of the U.S. Naval Academy who survived Adm. Hyman Rickover's brutal selection and training program and served briefly in the nuclear Navy. His father's untimely death forced him to abandon his nuclear career and return to the family's peanut farm near Americus, Georgia.

caused a major problem. Their spent fuel pools, which temporarily housed used fuel underwater as it awaited the expected transfer to reprocessing plants, now became de facto permanent storage sites.

Carter's energy advisor, former AEC chairman Jim Schlesinger,[125] retained the enthusiasm for monitored retrievable storage (MRS) that he developed at the AEC. Others disagreed, including some key officials at the new Nuclear Regulatory Commission. The NRC's William Bishop complained that above-ground storage put the commission "in the logically awkward position of being asked to license reactors that will produce wastes without assurance that the wastes can and will be disposed of safely." Nuclear utilities didn't much care about the final disposal of the spent fuel, as long as they could get it out of their glowing storage pools and off their books. Environmental groups, distrustful of Schlesinger and the industry, saw MRS as a way to preserve the reprocessing option, and wanted a final, geological burial site.

Many in the business of designing and building nuclear technology were unwilling to write off reprocessing. They were convinced that uranium was and would remain a limited resource, and that the next generation of nuclear power technology would be breeder reactors. Breeders require reprocessing to separate the plutonium fuel created by the operation of the reactors. So this group favored keeping the spent fuel, and its inventory of future fuel, available for future use.

The Carter administration, after an extensive interagency review process, offered what it saw as a compromise solution that came to be called away from reactor (AFR) storage. The federal government would take title to the waste from the utilities in return for a one-time fee. The energy

---

125 Schlesinger was the 1970s bi-partisan government utility infielder. With an economics PhD from Harvard, Schlesinger became an Office of Management and Budget official in 1969. President Nixon named him AEC chairman in 1971, at age forty, where he served until 1973, when he replaced Richard Helms as director of the Central Intelligence Agency for a brief, stormy six months. Nixon then appointed Schlesinger secretary of defense in 1973, where he served until November 1975, when President Ford fired him for insubordination. Schlesinger became Carter's energy advisor in 1976, no doubt harboring grudges against Ford. He helped Carter steer through Congress the legislation creating the Department of Energy, and Carter named him the nation's first energy secretary in October 1977.

department would then store the spent fuel rods in a temporary site—unlike the permanent site envisioned in the MRS concept—while the government developed a permanent, underground nuclear cemetery. The Carter plan envisioned a program to test and evaluate several geological sites for a final repository for both civilian and military waste, avoiding the rush to find a solution that sunk the Lyons enterprise.

Time was rapidly running out on the Carter administration, which was widely seen as a likely one-term presidency. By the time the Carter White House came out with its full views on nuclear waste in late 1979, Congress had already taken control of the action. The nation's next episode of nuclear waste frustration would start at the legislative end of Pennsylvania Avenue. *Science* reporter Luther Carter summarized this stage: "The recommendations by President Carter, coming in his last year in office when Carter was in a losing struggle for political survival, were given no priority and were not pushed by the administration on Capitol Hill. But in their emphasis on deep geological isolation as the first order of business and on investigating several potential repository sites thoroughly and then picking the best of them, Carter's recommendations did influence the legislation that eventually emerged."

The Carter legislative waste cotillion began in 1980, the final year of the Ninety-sixth Congress and the presidential election year. During 1980, both the House and the Senate passed nuclear waste bills, but a conference committee attempting to reconcile the two measures failed, in part because Congress was addressing other, higher-profile energy legislation.[126] The 1980 waste bill foundered on issues that would form the crux of the debate on the bill that eventually emerged from Congress in the Ninety-seventh Congress. The Senate bill, largely written by the Energy and Natural Resources Committee, included provisions for both MRS and

---

126 Congress, in the dying days of the Carter administration, created the U.S. Synthetic Fuels Corporation, a key Carter energy objective, which also proved to be a failure of the industrial policy paradigm. The quasi-governmental corporation, with an authorization of some $80 billion, was created to jump-start a private-sector effort to turn coal into oil and natural gas. Congress extinguished the program in 1985 as oil prices were falling and the synthetic fuels program floundering amid charges of mismanagement and corruption.

interim above-ground AFR storage.[127] The House bill, a product of the House Interior and Insular Affairs Committee, included neither Senate provision, but mandated permanent underground disposal, absent in the Senate bill.

Ronald Reagan easily defeated Carter in the 1980 election, taking office on January 21, 1981, and bringing an expressed interest in revitalizing nuclear power. The nuclear industry was demoralized by public opposition that was fed by the 1979 reactor meltdown at Three Mile Island, the still faltering economy, and the perception that there was no solution to the growing inventory of nuclear waste building up at the nation's nuclear power plant sites.

By 1982, utilities had cancelled seventy-seven reactor orders. Another eighteen would fall in 1982, fourteen more in 1983, and again fourteen in 1984. Joe Palladino,[128] the Penn State nuclear veteran whom Reagan named to head the Nuclear Regulatory Commission in 1981, told *Science* magazine's Carter in the summer of 1982, "If Congress passes the waste bill—that very much increases the confidence on the part of a number of the commissioners."

The Reagan landslide also brought a large number of Republican lawmakers to Washington. The GOP took control of the Senate by a 53–46 margin, the first time since 1954 that the Republicans had controlled either branch of Congress. Tennessee's Howard Baker, a nuclear enthusiast, replaced coal-state icon Robert Byrd of West Virginia as Senate Majority Leader.

Republicans gained thirty-four seats in the House of Representatives in the 1980 election, but that was not nearly enough to oust the Democrats from control. The Democrats controlled the House 243–192, and Tip O'Neil of Massachusetts remained Speaker of the House. Most important for the story of nuclear waste legislation, Rep. Morris Udall, a staunch

---

127 Parsing the differences between MRS and AFR is an exercise for nuke waste geeks and largely unproductive.

128 Nunzio Joseph Palladino (1906-1989) was a mechanical engineer who led the Westinghouse team that designed the reactor for the Nautilus. He joined the Penn State faculty in 1959 and became dean of engineering in 1966 until named to the NRC in 1981, where he served until 1986. He was a gentle and low-key leader at a difficult time for the NRC, leading its often incoherent response to the 1979 accident at Three Mile Island.

liberal and environmentalist from Tucson, Arizona, brother of former Interior Secretary Stewart Udall, remained chairman of the House Interior and Insular Affairs Committee.

While the White House professed great interest in passing waste legislation, the Reagan administration didn't play much of a role in the debate over the 1982 legislation, preferring to let its forces in Congress do the heavy political lifting. That was largely because the issue didn't matter very much to the major players in the White House, including the president. They were focused on more important issues, including the existential struggle with the Soviet Union.

With decontrol of federal oil prices now fully in place under Reagan (having begun under Carter), oil prices were falling dramatically, gasoline prices were following, and energy was quickly receding from the political agenda. The Energy Department, created with much fanfare in the Carter administration, had proved to be a largely irrelevant institution. Reagan's first energy secretary, Jim Edwards, was a former South Carolina dentist and accidental politician who served one term as Republican governor in the Palmetto State. His most memorable contribution at DOE was his comment, not long into his sole year at the Forrestal Building, "I can't wait to get my hands back in spit."

The Reagan administration in October 1981 issued what it billed as a major policy statement on nuclear power. Reagan was elsewhere occupied, and Edwards delivered the address—and the subliminal message that the White House really didn't give a damn about nuclear power. The message "repealed" the Carter ban on reprocessing, with no tangible results, and called on DOE and the industry to solve the nuclear waste conundrum—political humbug.

The nuclear waste legislative debate kicked off in the Senate, where Idaho's Jim McClure was the chairman of the Senate Energy and Natural Resources Committee. McClure was a dedicated advocate of all things nuclear.[129] He worked closely with the leading Democrat on the committee, J. Bennett Johnston of Louisiana, another dedicated advocate of all things nuclear, who was particularly interested in assuring that the waste would

---

129 McClure was particularly interested in moving military nuclear waste out of Idaho, where it was sitting in above-ground tanks at the federal weapons program's Idaho lab.

not end up stored in the extensive salt deposits in his home state. McClure also coordinated his legislative efforts with Sen. Alan K. Simpson, the Wyoming Republican and nuclear agnostic, most interested in advancing the interests of Wyoming uranium mining companies. Simpson chaired the Senate Environment and Public Works Committee.

Getting legislation through the Senate was relatively straightforward. The energy committee has never been a locus of partisan or ideological contention. It has always been dominated by energy interests; most of the legislation it has dealt with over the years has involved parceling out provisions and conditions aimed at the economic interests of its members. For many years, service on the committee was much in demand among senators of both parties in election years, as its ties to energy industry interests have yielded generous campaign contributions.

A major legislative hang-up in the Senate was homeland security—that is, the Senators wanted legislative assurances that their homelands would not be burdened with disposing of the waste. Stealing from a comment Sen. Russell Long (D-La.) once made about taxes, the attitude of the senators about nuclear waste was, "Don't dump it on you; don't dump it on me. Dump it on the guy behind the tree."

The Senate legislation finessed the issue of where the waste would be dumped by providing for two final sites, potentially spreading the pain so that no one state could say it was being singled out for the short, radioactive straw. The West would get the first site, which the legislators expected would be sited in salt or stable volcanic rock. Many observers also expected that the first site would be located on land the federal government already owned, such as at Hanford or on the Nevada Test Site. The East or upper Midwest would get the black spot for a second dump, probably located in granite and on private land.

The Senate bill also contained provisions for both MRS and AFR storage, at Johnston's insistence. This provided further political insulation against storing the waste in Louisiana, even though many technical experts still believed salt represented the best storage medium—the problems in Kansas notwithstanding. The final Senate bill, passed in April 1982 by a 69–9 vote, provided a mechanism for a state or Indian tribe to object to siting of the waste dump. The aggrieved party could file a petition with

Congress that would trigger a vote. If either the House or the Senate agreed with the state or tribe, the veto would stand.

The Senate bill also contained a funding mechanism for the waste program. Nuclear utilities—that is, their customers—would pay a fee of a mill (a tenth of a penny) for each kilowatt-hour of nuclear generation, or about $5 million per reactor each year, into a special fund to be used for the waste program.[130] Along with the fee, the legislation provided for an aggressive schedule, one that many experts found far too optimistic. DOE would characterize at least three geological sites and recommend one for the first repository by January 1, 1986. DOE would recommend a second site by January 1, 1989. The NRC would be required to rule on a DOE license application for the first site by the end of 1989 and the second by the end of 1992. The federal government would take title to the utility spent fuel by the end of 1998. None of the dates were remotely realistic.

The real contention on waste legislation came in the House, where the industry and its Republican allies had considerably less political traction. At the time the Senate passed its bill in the spring of 1982, the House legislation was mired in committee hearings and markups. Four House committees had significant jurisdiction—Morris Udall's Interior committee, John Dingell's Energy committee, the Science committee, and the Armed Services committee. The Rules committee, under the direction of Missouri liberal Democrat Richard Bolling, would have to tease out the jurisdictional tangle before the House could vote on a final bill.

By fall, with the congressional clock ticking down on the Ninety-seventh Congress, the House appeared deadlocked. It was up to Bolling to break the legislative logjam. Bolling told Luther Carter, covering the bill for *Science,* that he held both the environmental community and the industry to blame for the impasse. "The environmentalists want it perfect and would like to kill nuclear power," Bolling said. "The industry wants what will serve the interests of the stockholders, and to hell with everything

---

130 Like all such fees, the money actually flowed into the general U.S. Treasury, where it would be mingled with all other federal dollars and not kept in a dedicated federal "lockbox."

else." Bolling turned to Udall[131], a lanky, funny, self-deprecating former pro basketball player and lapsed Mormon, a beloved figure in the House, to work a deal.

Udall soon got together with Dingell and his relevant subcommittee chairman Richard Ottinger, a New York liberal and environmentalist, then with the House Republicans, reaching a preliminary compromise. Udall's approach adopted the spread-the-pain strategy of the Senate legislation, proposing sequential searches for two sites, starting in the West and following with a second repository in the East or Midwest. The negotiators agreed to a series of new provisions that would give the states more options to protect themselves from the DOE dump, along with provisions for the MRS and AFR storage as a backstop. By late September, Bolling's Rules committee cleared a compromise bill—H.R. 7187—for floor action sometime that fall.

But the environmental groups, whom Udall and Dingell thought had been bought off with improved state veto provisions and additional language on environmental review of candidate sites, jumped off the legislative ship. In late November, the enviros notified the House leadership that they would oppose the compromise bill. Their main objections were to the above-ground options, which they saw as undermining their goal of geological disposal.

With time running out—Congress had to adjourn by the end of the year or start all over again with a fresh Congress—the environmental groups were now able to largely drive the process. During the final House negotiations, two savvy environmental lobbyists, Brooks Yeager for the Sierra Club and David Berick for the Environmental Policy Institute, worked a range of House Democrats[132] to plant what they described during the closing days of the Congress as "time bombs" in the legislation.

The House took up the waste bill for final floor passage on November 29, 1982. It passed by voice vote after three days of debate. The major amendment the House adopted on the floor came from Udall, as a sop to

---

131 Two Udalls, Mo's son Mark, from Colorado, and his nephew, Stuart's son Tom, of New Mexico, served together as Democrats in the U.S. Senate starting in 2009.

132 Yeager and Berick often worked with staffers for unlikely members such as G.V. Sonny Montgomery, a conservative Mississippian who frequently denounced "that goddamn Sahara Club," but represented a district with geologically-attractive salt deposits.

the environmentalists. It stretched out the schedule for a DOE study of MRS to five years from the two years in the legislation and one year in the Senate bill. The House legislation also limited the terms for temporary AFR storage. The House also rejected an absolute veto by a state or Indian tribe, keeping language that tracked the Senate's state veto.

House passage set the stage for a frenzied House-Senate conference committee that largely worked behind the scenes as 1983 approached. The conferees largely split the differences between the two bills, as most House-Senate conference committees do. DOE would study five candidate sites, recommending three to the president for a first-round selection for characterization by January 1, 1985. The president would pick the finalist by March 31, 1987, and the NRC would have three years to license the site.

In the meantime, DOE would sift through another five sites, recommending three to the president for a second dump by July 1, 1989, with the president's pick coming by March 31, 1990. DOE would take possession of the spent fuel by December 31, 1998. The program would be funded by the mill per kilowatt-hour fee. The compromise contained authorization for a temporary above-ground, monitored storage site, but that would have to be separately authorized by Congress.

A filibuster threat by Wisconsin Democratic Senator William Proxmire proved to be the only stumbling block. Wisconsin housed a candidate site. Proxmire offered an amendment requiring both houses of Congress to act to overturn a state or Indian tribe veto, which won the support of governors from New Mexico, Nevada, Utah, and Washington. Fearing the Proxmire talk-a-thon, the conferees accepted the amendment. The Senate approved the final bill by voice vote on December 20, 1982. The House followed with a 256–32 vote hours later. The 1982 Nuclear Waste Policy Act became law.

For the first time, the United States had a formal nuclear waste disposal policy, focused on geological disposal. The question became whether the new law would work. It didn't take long to discover that the answer was no.

The most perceptive assessment came from Eliot Marshall, another *Science* reporter, who wrote, "A bill like this would have to be considered only a hesitant first try at solving the nuclear waste problem. It deals with none of the technical disputes and leaves the highly difficult task of site selection to the bureaucracy."

23. Screw Nevada and Nevada Will Screw You

John Stewart Herrington was an unlikely energy secretary, at least as unlikely as dentist Jim Edwards, the Reagan administration's first energy chief.

A short, pugnacious former Golden Gloves boxer, motorcyclist, and erstwhile Marine lieutenant, Herrington came to Washington from a job as a California business lawyer with two aims: to further the programs of Ronald Reagan and advance the future of the Republican party. A 1983 *New York Times* profile characterized him as a "two hundred–proof Reaganite."

Herrington, a Los Angeles native born in 1939, graduated from law school at the University of California at Berkeley's Hastings Law School in 1964 and in 1966 became a deputy district attorney for Ventura County. He also became a volunteer precinct organizer in Ronald Reagan's successful 1966 gubernatorial campaign. His ardor for the former actor never waned.

When Reagan was elected president in 1980, he brought Herrington, who had served as an advance man during the campaign, with him to Washington as assistant Navy secretary in charge of manpower. Two years later, as the administration began thinking about the 1984 reelection campaign, Herrington moved to the West Wing of the White House, as presidential personnel director and troubleshooter.

In early 1985, with the reelection campaign successfully behind him, Reagan announced a major cabinet shakeup. Energy Secretary Don Hodel, who was an energy expert, was moved back to the Interior Department, succeeding William Clark as secretary. Hodel had come to the Reagan administration as second in command at Interior to the controversial James

Watt. Reagan named Herrington, who had no energy experience whatsoever, to succeed Hodel at the energy agency.

The nomination was a surprise. Brooks Yeager, the Washington lobbyist for the Sierra Club, who had been influential during the passage of the
1982 Nuclear Waste Policy Act, proclaimed, "The guy is a cipher. As far as
we can find out, he doesn't have the remotest qualifications or background
in energy management." That was not an overstatement.

At his Senate confirmation hearing before a friendly, Republican-
dominated energy committee, Herrington deflected substantive questions
about energy policy and energy issues. Many of the tougher shots were
thrown by committee Democrats using the hearing to warn the nominee
that they knew a heck of a lot more than he did and they would be testing
his sure-footedness at the agency.[133]

As Herrington settled into his expansive digs at 1000 Independence
Avenue, his agency began the process of implementing the new waste law.
That process rested on a series of hearings aimed at winnowing down the
first-round candidate sites into the three the agency was required to present
to the president in two years.

The hearings were a warning that the department was in serious trouble. In a scholarly article in 2000, American University political scientist
James Thurber noted the "public dissatisfaction that existed with the [new
waste law's] implementation." Thurber observed that as the first series of

---

133 I reported on Herrington's confirmation hearing for Energy Daily, employing box-
ing images. I had him bobbing and weaving, deflecting punches, and concluded that
the Democrats hadn't laid a glove on him. The Republican press spokeswoman for the
committee was ecstatic that I had treated Herrington so well. The next day, I got a phone
call from the White House that Mr. Herrington wished to see me in his office the next
morning. Thinking he wanted to offer me a job, I showed up bright and early at the North
Gate. After letting me cool my heels for an hour-and-a-half, a nice young woman who
worked for Herrington showed up and escorted me to his windowless West Wing office,
painted yellow and decorated in Government Services Administration faux-Colonial furniture. Herrington proceeded to dress me down, saying how angry he was about my article
and how much better the (hagiographic) report in the LA Times captured his performance. When I got over the shock, I reminded Mr. Herrington that I had been covering the
energy department long before he arrived in Washington and I expected to be covering it
long after he left. We parted with an icy handshake, but he treated me no differently than
any other reporter (which is to say, guardedly) during his term.

public hearings began, "the scope of the conflict expanded over the nuclear waste issue as NIMBY sentiment spread. The wider the conflict, the less influence the congressional nuclear power advocacy committees and the DOE had over the outcome of nuclear waste policy."

Thurber added, "The DOE's handling of the public hearings process worsened the situation. DOE personnel believed in an organizational culture and norms that ignored politics and refused to discuss issues except in the technical jargon of the engineering profession. To the DOE, site selection was not political but technical. The DOE held hearings as mandated by [the waste law], but its attitude toward the hearings and the ways in which it responded to the public made its political problems worse."

The DOE grunts weren't political. John Herrington was political to the bone. And the politics of the impending decisions on selecting a waste dump were politically radioactive. The first-round choice was the least of the problems. The three sites that looked most promising were in Deaf Smith County, Texas, at the Hanford nuclear weapons site, and at Yucca Mountain on the Nevada Test Site. Most careful observers expected either Hanford or Yucca Mountain as the choice, as both were located on federal land. The smart money was betting on Yucca Mountain, as Washington and Oregon both had incumbent Republican Senators up for reelection (Slade Gorton in Washington and Bob Packwood in Oregon). Nevada was familiar with the nuclear enterprise and was politically impotent in Washington. The state's senior senator, Reagan's good friend Paul Laxalt, was retiring, and former Democratic member of the House turned Republican Jim Santini was expected to fill his seat in the Senate. Whoever the Nevadans elected would be a powerless freshman.

Herrington was deeply concerned about the 1986 elections, as the Democrats were poised to make solid gains typical of non-presidential election years. The GOP had gained control of the Senate in the 1980 elections and held on in 1982 and 1984. But they had a number of contested races coming in 1986, many in eastern and mid-western states that would be candidates for the second nuclear waste dump.

In a shocking announcement on May 28, 1986, Herrington abruptly dumped the second round. He announced that the three expected sites in Texas, Washington, and Nevada would go to the president for a final choice. The seventeen states that were facing the honor of hosting the second site would be spared as Herrington put the process on bureaucratic ice. Nuclear reactors were not being built at the pace expected, he said, so there was no urgency to find a second repository. "Based on the progress we have made toward selecting a first repository site," Herrington said at a Washington press conference, "I have reassessed the timing of the department's activities toward identification of candidates for a second repository, and I have decided to postpone indefinitely plans for any site-specific work."

Herrington's announcement amounted to a death sentence for the 1982 law, exposing its fatal flaw. Spreading the pain did not result in a sense of shared sacrifice, but simply increased the political stakes in the siting decision. The "time bombs" hidden in the law exploded with the first attempt to implement this flawed approach to an intractable problem. Political scientist Thurber commented that as the first real deadline approached, "outcry was so loud that for many in Congress, opposition to the DOE's search for a repository had become a political necessity and few legislators were willing to risk their careers" on the waste program.

Shortly after Herrington's announcement, Mo Udall commented, "The program is in ruins and our goal of siting a repository seems further away than ever." He was right.

Immediately after Herrington's announcement, the first-round states of Texas, Washington, and Nevada sued. In Texas, where the potential repository was on private land, state officials refused to issue a permit necessary to move forward with site characterization.

That fall, Congress dramatically cut funding for the waste program. Luther Carter observed, "The irony was that while stopping the second round was probably politically inevitable, this decision made the already bad problems of the first round even worse." The first-round states were adamant that they would not share the burden alone. Washington's Democratic governor Booth Gardner testified in Congress, "If the federal government won't play by the rules, we will see you in court. The future of the repository will be tangled in the nation's court system for years to come."

The fall elections were a disaster for the Republicans, as Herrington had feared. The Democrats picked up eight Senate seats and now controlled the institution by a 55–45 margin, with coal-state titan Robert Byrd installed as Majority Leader. In Washington State, voters unseated Republican

Senator Slade Gorton and narrowly elected Democrat Brock Adams.[134] Gorton had been a strong proponent of the second round, hoping to deflect attention from the site in Washington. In North Carolina, Democrat Terry Sanford ousted Jim Broyhill, who had been a key Republican House player in the 1982 act and was appointed to the Senate when Republican John East committed suicide in 1986. In Nevada, a nondescript member of the House, Harry Reid, won an open Senate seat, knocking off Republican nominee Jim Santini with only 50 percent of the vote, as a Libertarian candidate took almost 2 percent, most of it from Santini.

The impending collapse of the program was obvious. In January 1987, the Reagan administration announced a five-year delay in the schedule for the first repository. DOE issued a new schedule for the site selection, citing the technical and political difficulties that lay behind the decision. The agency also noted the congressional action to cut funds as requiring a stretching out of the schedule. The administration also said it would store power plant wastes temporarily at Oak Ridge, although it had no legal authority for that action and soon was forced to back off on that part of the plan.[135]

Fearing a full-scale replay of the legislative chaos that led to the 1982 act, Louisiana Democratic Senator J. Bennett Johnston, by then chairman of the Senate Energy Committee, began looking for a way to get the waste express back on the tracks, and not headed toward the attractive salt beds in his home state. He turned to an idea that journalist Luther Carter had hatched in *Science* magazine: target Yucca Mountain and forget the rest of the process.

The idea was simple. Yucca Mountain's volcanic tuff looked like a good place to dump the nuke waste. It appeared to be dry and stable and securely located on the Nevada Test Site, where it was out of the reach of

---

134 One of the first hires Adams made for his staff was David Berick, who had lobbied for environmental groups during the passage of the 1982 waste act and was now working as a lobbyist for the Union of Concerned Scientists.

135 John Herrington served out the second Reagan term as energy secretary; then returned to California. He served as chairman of San Diego-based publishing company Harcourt, Brace, Jovanovich and was chairman of the California Republican Party as the state became almost entirely controlled by Democrats. He currently owns a steak house in Walnut Creek, California.

state officials. What's more, Nevada didn't pull much political weight in Washington. The state's senior senator was Republican Mayer Jacob "Chic" Hecht, in his first, and only, term. He was a low profile, malapropism-prone Reaganite who never carried any heft in Washington. Influential Republican senator Paul Laxalt decided not to run for reelection in 1986, contemplating running for the GOP presidential nomination in 1988. The state's two members of the House of Representatives were Democrat James Bilbray and Republican Barbara Vucanovich, household names only in their own households.

Nevada's very junior senator was Harry Reid, who just turned forty-six when he was sworn into office in January 1987.[136] Reid was a slight, mild-mannered Mormon born in the tiny mining community of Searchlight, Nevada. He was that kind of individual who could easily melt undetected into a crowd. If Harry Reid walked into a Washington cocktail party with a name tag on in 1987, he likely would not have been recognized. But Reid had many virtues, chief among them persistence.

Reid's political career was mixed before he narrowly won the 1986 Senate race, but demonstrated his willingness to get up and keep fighting after a defeat. After earning a law degree from George Washington University, while working for the U.S. Capitol Police, Reid returned to Nevada. In his twenties, he served as city attorney for Henderson, the state's second largest city and part of the Las Vegas metro area. In 1970, as a thirty-year-old, Reid was elected lieutenant governor to Democrat Mike O'Callaghan, his political mentor and his former high school teacher.

In 1974, Reid ran for the U.S. Senate and lost narrowly to Republican Laxalt. The next year, he lost a race to be Las Vegas mayor. After the 1980 census, fast-growing Nevada gained a seat in the U.S. House of Representatives. Reid ran for the new seat and won. He served two terms before making another run for the Senate in 1986 and winning unexpectedly. Reid throughout his career opposed the siting of the nation's nuclear waste dump in Nevada and he cruised to reelection in 1992; won again, narrowly, in 1998; won easily in 2004, and survived a close race in 2010, targeted by Tea Party Republicans.

---

136 He was six months younger than energy secretary John Herrington. Both had been amateur boxers as young men.

Luther Carter in his writings in *Science* magazine and in his 1987 book, *Nuclear Imperatives and Public Trust*, had suggested settling on Yucca Mountain as the solution to the radioactive waste morass. "Unless a confident show of progress is made soon," he wrote in 1987, "the geological disposal effort will take on the appearance, if indeed it has not done so already, of an interminable trek toward an ever-receding mirage."

Carter suggested that Yucca Mountain had good potential as a nuclear waste dump. But Carter did not suggest some sort of slipshod, kangaroo court approach to designating Yucca Mountain as the final repository site.

That was up to Bennett Johnston and his partner on the Senate Energy Committee, Idaho Republican Jim McClure. Johnston was not only chairman of the energy committee but also chairman of the powerful Senate energy appropriations subcommittee. Together, Johnston and McClure hatched a plan that would direct the Department of Energy to drop the Hanford and Deaf Smith sites, requiring the agency to characterize only Yucca Mountain. They worked to have the language inserted into an omnibus 1987 budget bill, a must-pass measure, during a House-Senate conference committee, so the Johnston and McClure plan was never subjected to debate in any congressional committee or on the floor of either the House or the Senate.

When the bill went to Reagan for signature, Johnston bragged, "I think it's fair to say we've solved the nuclear waste problem with this legislation." Reid complained it was the "Screw Nevada bill." Johnston was wrong, and Nevada and Reid eventually screwed those who decided to bully the state by imposing the nation's nuclear waste on them.

It took time, but patience was Harry Reid's trump card, and time favored Nevada. The passage of time aided both the politics of opposition to Yucca Mountain and the scientific case against the site, originally thought to be a technical slam dunk for disposal.

The political case for geological disposal of spent nuclear fuel hinged on the fact that the nation's nuclear utilities would run out of space at their reactors to store the fuel and that a new round of nuclear construction would add to the growing burden of used fuel rods. But neither came to pass in the fullness of time. Utilities were able to rearrange the spent fuel in the radioactive swimming pools where they initially cooled down after leaving the reactor. More importantly, the industry developed large concrete and steel, air-cooled caskets for housing the spent fuel on site. Every couple of years, the utilities and the nuclear power industry would proclaim an imminent shortage of storage space at the reactor site, which never seemed to materialize.

And new reactors did not appear on American soil. The predicted second wave of nuclear development never materialized. Conditions never seemed to be quite right for new nuclear plants. In the 1980s and early 1990s, decontrol of natural gas prices in the United States produced a long-lasting glut, the "gas bubble" that wasn't a bubble at all but a decade-long oversupply, which led to a construction boom for gas-fired generating capacity.

The same time period saw the rise of non-utility generators who did not have captive customers upon whom they could load up the high, upfront capital costs of building nuclear plants, even though the nukes had extremely low fuel costs and favorable environmental impacts on the air. The new non-utility generators, or NUGs, as the industry called them, built plants that were essentially jet engines, burning gas. These combustion turbines generated electricity directly. As the demand for electricity grew and firmed up, the generating companies captured the heat from the jet engines, used it to make steam, and ran the steam through steam-turbine generators, yielding more cheap electricity.

As time passed, the Nevadans and their allies, including some anti-nuclear groups, mounted a delaying action in Congress and the courts. They made the case that Congress had acted precipitously, picking the site long before anyone had the science in hand to say that Yucca Mountain was even an acceptable place to put radioactive waste, let alone the best place.

It became clear that Nevada's technical and political objections were taking hold. In 1989, just two years after Congress passed Nevada the black spot, DOE again pushed back the date at which it could have a repository ready, from 2003 to 2010. As a temporizing measure to deal with that delay, the NRC gave utilities a green light for storing waste onsite, outside of the storage pools in the newly-developed storage casks, ruling that this was a safe way of handling the waste.

A recent history of the waste program, *Fuel Cycle to Nowhere* by environmental lawyers Richard Burleson Stewart and Jane Bloom Stewart, noted that the technical issues surrounding Yucca Mountain continually surprised the supporters of the project. Those technical surprises "produced repeated changes in DOE's technical assessments for Yucca and revisions to the project design and objectives. The emergence of additional and more complete understanding spawned skepticism about Yucca's suitability, and not just from hard-line opponents."

The more DOE examined the site, not just through core samples, but from widespread and deep excavation, the more complex and ambiguous the assessment of the suitability of the site became. A crucial issue was the ability of water to penetrate the volcanic tuff that made up the geology of the mountain and the ability of the water table to penetrate the portion of the mountain where the waste would be buried. The assumption from the beginning—and one of the reasons Johnston was able to express complete assurance that doubling down the government's bet on Yucca Mountain was a sound move—was that the site would be dry.

One of the scientists examining the site was Victor Gilinsky, a nuclear energy old-timer who was one of the earliest members of the U.S. Nuclear Regulatory Commission when it was created in the breakup of the Atomic Energy Commission. In 2002, Gilinsky, then on the Cal Tech faculty, visited the test tunnel at Yucca Mountain as a consultant for the state of Nevada and told a congressional committee, "I was pretty surprised therefore to find myself standing in the middle of this mountain with water dripping out of it and hitting me on the head." Water in the repository area, Gilinsky noted, "Will cause corrosion and fissures in the nuclear waste containers and if that happens, the containers could easily start leaking, distributing their contents into the local ecosystem."

When the Department of Energy first started to understand Yucca Mountain's major water problem, the government engineers and program managers responded by changing the rules of atomic waste disposal, deciding that it wasn't so important that the area remain dry. Instead, DOE would design novel metal containers—what some snarky skeptics called "miracle metal"—to hold the waste and isolate it from the water in the mountain.

John Bartlett, at one time the highly-respected director of the DOE waste disposal program, told a federal court that "rates of water infiltration into the mountain were on the order of [one hundred] times higher than had been expected; that water flowed very rapidly through fracture pathways in some of the geologic layers (like flow through a pipe rather than dispersed flow through a medium like a bed of sand); and there appeared to be unexpected 'fast pathways' for movement of radioactivity from the repository to the water table about one thousand feet beneath it." So DOE came up with what it called "engineered barriers"—high tech tin cans. Bartlett described the performance of this new approach as "unknowable."

As the science and engineering supporting the choice of Yucca Mountain was weakening, Nevada senator Harry Reid was gaining strength and influence. While unprepossessing, Reid was demonstrating over the years that he was adept at correctly counting votes in a body where transparency is most often an illusion. Reid also showed a skill at understanding the often murky motives that moved his fellow senators' policy positions.

After the reelection to his third Senate term in 1998, when he defeated Rep. John Ensign[137] by four hundred votes, Reid's Democratic colleagues elected him as party whip, an important job with the responsibility of lining up votes for the party leader, who was South Dakota's Tom Daschle. Daschle and Reid made an effective team during the early years of the twenty-first century, whether the Democrats were in the minority or majority. Daschle was "Mr. Outside," good looking and well spoken, a man who moved easily among Washington political and power elites. Reid was "Mr. Inside," a rumpled presence who worked the corridors and cubbyholes of the Senate stroking egos, twisting arms, and lining up votes for the Democrats.

Reid demonstrated his party loyalty in 2001, after Republican gains in the 2000 election gave them one-vote control of the Senate. The Democrats began wooing Vermont's wobbly Republican senator Jim Jeffords, who was uncomfortable with the rightward direction of the GOP. If Jeffords left the party, Democrats would regain control and Daschle would be majority leader and control the Senate's business. Reid would be in line to chair the Senate Environment and Public Works Committee, with a major piece of jurisdiction over the DOE waste program.

Jeffords, a strong environmentalist, was the most senior Republican on the environment committee, although the Republican caucus refused

---

137 Ensign later won a Senate seat in 2000 and was seen by some as a potential GOP presidential candidate. He was forced to resign in 2011 as a result of a sex scandal.

to allow him to become chairman because of his party unreliability and green policy coloration. Reid offered Jeffords a deal, proposing that the Vermonter become committee chairman in a Democratic Senate, giving up his own claim to the chair. Jeffords agreed, Daschle became Senate Majority Leader, and Harry Reid remained Daschle's top lieutenant as majority whip.

Reid also took control of a major money dispenser in the Senate, the energy and water appropriations subcommittee of the Appropriations Committee, the same subcommittee Bennett Johnston had once ruled. From there, Reid could control the funding for Yucca Mountain. While the 1982 law set up the nuclear waste fund, Congress from the start treated it as just another pot of money. Over the years, as customers of electric companies with nuclear plants contributed their one mill per kilowatt-hour of electricity to the government, Congress consistently cut the budget for the waste program, spending the money on other government programs. Reid continued clamping down on Yucca funding, strangling the project's finances as the science undermined its rationale.

In the 2004 national election, when George W. Bush was easily elected to a second term, Tom Daschle—to most everyone's surprise—lost his South Dakota seat. But the Democrats kept control of the Senate, and Harry Reid became the Senate Majority Leader. By most estimates, Yucca Mountain was already a dead dump walking by 2005, but Reid's ascent to the top job in the U.S. Senate sealed its fate.

Also in 2005, a young freshman Democratic senator from Illinois with the improbable name of Barack Hussein Obama approached his majority leader and pledged his support for killing the Nevada waste dump. Obama knew that if he helped Reid in the Senate, then Reid would help him navigate an institution whose rules and mores were unfamiliar. Obama was also basking in the acclaim from his electrifying speech to the 2004 Democratic National Convention and surprising Senate election, thinking that maybe he could run for president someday.

By 2007, following the 2006 by-year elections where Democrats recaptured the House of Representatives and kept control of the Senate, Obama was formulating his plans for a run for the Democratic presidential nomination. Obama quickly recognized that Nevada's January Democratic caucus, the third major political event in the nation in early 2008, would be important in giving him credibility and political momentum. In a

2007 letter to the *Las Vegas Review Journal*, Obama said, "I have always opposed using Yucca Mountain as a nuclear waste repository." Joining all the other Democratic presidential candidates, Obama pledged to kill Yucca Mountain as a repository if elected.

When Obama easily won the election, he moved quickly to redeem his promise to Reid, whom he would need to get his legislative package through Congress. Carrying out Obama's promise, Energy secretary Steven Chu in May 2009 withdrew the administration's support for the project. The congressional Democrats followed by slashing funding for the program even below where it had been in the Bush administration. The Nuclear Regulatory Commission, under the direction of Gregory Jaczko, a PhD nuclear physicist who had been Reid's scientific advisor before Bush named him to a Democratic seat at the NRC, began shutting down the agency's licensing process for the waste dump.

The conclusion of environmental lawyers John and Jane Stewart stands as an epitaph for sixty years of U.S. failure to manage nuclear waste: "The nation again confronts its legacy of highly radioactive defense and civilian waste, this time without a plan or even—under current statutes—legal authority to develop a repository or a consolidated facility."

Epilogue: Some dumb ideas never die

The Manhattan Project was a marvelous engineering success, but only on a limited front. It successfully captured the power of the atom for the purpose of war and mass destruction. Its success blinded postwar policymakers in the mid-twentieth century, who came to believe that big science melded to big government was the path to scientific progress.

The Manhattan Project proved to be a false god. As this book argues, the big-science approach to public science and engineering—the very model of what has come to be known as industrial policy and mimicked around the world—led to feckless, wasteful, needless dead-ends. We poured our wealth into bombers that could never fly, construction projects that could never work, mining and stockpiling fuel that we didn't need, technologies that never delivered, and waste disposal projects that gave unintended demonstrations of the meaning of the word "waste." Around it all, we developed an administrative and bureaucratic edifice that distorted our politics and misled our leaders and our people.

The folks who brought us these follies and failures weren't malefactors by any sensible definition (including the popular villain Edward Teller). With good motives, exceptional educations, broad experience, and true public spirit, they were mostly brilliant. But they were often blinded by their success and pride in their accomplishments. On top of that, there was a pervasive attitude that money didn't matter in the pursuit of atomic energy. British Energy minister Charles Hendry, in late 2011, summarized things well. Speaking to Britain's Royal Society, he said that in the postwar period, government energy agencies operated "like an expense account dinner: everybody ordering the most expensive items on the menu, because someone else was paying the bill."

While we should laud the virtues and many of the fruits of the labor of the leaders of the past, we should also see clearly where they failed and, as best we can discern, why.

Economist Lawrence Summers, adviser to Democrats from Carter to Obama, said recently, in the context of the Obama administration's failed investment in a dodgy solar energy project, that the government is a "crappy venture capitalist." That has been true for a very long time.

The spirit, concepts, and hubris flowing from the Manhattan Project have remained with us up to today. Hardly a month has passed without figures in policy and political circles proclaiming loudly, with nary a hint of doubt, that what America or the world needs is a new Manhattan Project to revitalize the economy, save the environment, or stretch our reach into outer space.

That's just what we don't need, but some dumb ideas never die.

Fly It

In 2008, a collaboration among the UK aerospace industry, two British universities, and the British government—known as the Omega Project— floated the idea of resurrecting nuclear-powered airplanes.

The need for flying nukes, explained Ian Poll, a professor of aerospace engineering at Cranfield University, flows from the specter of man-made global warming. In a 2008 interview with the *Times* of London, Poll, a distinguished engineer for many years, said, "We need a design which is not kerosene-powered, and I think nuclear-powered aeroplanes are the answer beyond 2050. The idea was proved fifty years ago, but I accept it would take about thirty years to persuade the public of the need to fly on them."

Poll's grasp of history may have been a bit uncertain, given that the idea was never proved, but his vision of the future was firm, if familiar. In his formulation, the atomic airplane—not a bomber this time, since the Cold War ended some twenty years ago, but a passenger vessel—would fly nonstop from London to Sidney or Auckland. There would be zero pollution—under an unstated assumption of no uncontrolled radioactivity.

What about shielding the crew and passengers from the local radiation of the engines? Not a big deal, Poll told the *Times*. "It's done on nuclear submarines and could be achieved on aircraft by locating the reactors with the engines out on the wings. The risk of reactors cracking open in a crash could be reduced by jettisoning them before impact and bringing them down with parachutes." Poll called for a large, government-funded program to develop the new generation of atomic-powered flight.

Poll's A-plane would have to be large, at least twice the size of a Boeing 747 by several estimates. It would require new airports with new landing strips, and docking stations miles from existing terminals. The trip from the plane to the terminal would seem interminable to many passengers after a long flight.

Because of the local radiation, pilots could only fly for a limited time before exceeding radiation limits.[138] Frequently fliers might also have to limit their flights on the A-planes to avoid exceeding radiation limits. Nuclear passenger planes, says Theodore Rockwell, a veteran radiation expert at the Oak Ridge National Laboratory, are "not good for anybody."

One of the most remarkable aspects of Professor Poll's paranormal vision is that it took place years after Arab terrorists crashed two conventional jet passenger planes into the World Trade Center in New York, demolishing them. In a 2008 article in *Scientific American*, David Lochbaum, a nuclear engineer at the U.S.-based Union of Concerned Scientists, scratched his head and said, "We've been worried since 9/11 about how to protect against bad guys hijacking an aircraft and crashing it into a nuclear power plant upwind of a heavily-populated area. Let's now put the nuclear reactor in the plane itself, so they can target cities without a nuclear plant upwind?"

Lochbaum called Poll's idea "a Christmas present for the terrorists of the world."

Fortunately, Poll's attempt to revive atomic flight has failed—at least for the time being. The aptly-named Omega Project[139] had UK government funding from 2007 to 2009, with a mission of looking at the environmental implications of commercial aviation (Poll posed the notion of atomic

---

138 Because the planes would flight at high altitudes, the background radiation is already considerably higher than background at sea level.

139 Omega is the twenty-fourth and last letter in the Greek alphabet. It is often used to signify the end or limit of a set or series, with alpha, the first Greek letter, as the beginning. In the Christian New Testament's book of Revelation, the final book, God is described as the "alpha and omega, the beginning and the end, the first and the last" (22:13, King James Version).

flight on his own) and organized by Manchester Metropolitan University. After a series of worthy academic tomes, none of which appear to have had any discernible impact on any field of inquiry, the group quietly passed into the mists of history.

Since then, however, there have been whispers and murmurs about powering the latest, highest-tech war planes—the remotely piloted drones the United States is using widely in the wilds of Pakistan and Afghanistan—with small nuclear reactors. This, of course, gives new meaning to the term "collateral" civilian casualties.

Blow it up

While the United States abandoned its "peaceful nuclear explosions" ambitions over thirty-five years ago, and the Russian remnants of the Soviet Union seem to have no interest in rearranging the physical landscape with hydrogen bombs, China appears to have big plans for blowing up portions of the Himalayas to reroute a major river system.

Around 2003, China began publicly discussing the idea of rerouting the Brahmaputra River and its tributaries, which begin in Tibet as the Yarlung Tsangpo River and flow into India in the state of Arunachal Pradesh, through territory that is the subject of dispute between India and China, into the Indian state of Assam, and then into Bangladesh. The flood-prone river joins the lower Ganges and empties into the Bay of Bengal through a giant delta. It is one of Asia's most important, and least polluted, rivers.

For some twenty years, rumors have circulated that China planned to dam the river in Tibet, diverting its flow to China's desert regions as well as generating electric power for China's burgeoning industries. China has already built a dozen dams on the river in Tibet, without consulting its downstream neighbors. The rumored concept is that China would divert the flow of the Brahmaputra into China's Yellow River basin, watering the Chinese desert and impoverishing India and Bangladesh. Many Asian analysts say water will be the key natural resource in the future, defining the course of economic development in the big rivals, China and India.

Indian geopolitical analyst Brahma Chellaney at the Center for Policy Research in New Delhi wrote in 2009 that "China is now pursuing major inter-basin and inter-river water transfer projects on the Tibetan plateau, which threatens to diminish international river flows into India and other co-riparian states." Chellaney noted, "As its power grows, China seems determined to choke off Asian competitors, a tendency reflected in its hardening stance toward India…Water is becoming a key security issue in Sino-Indian relations and a potential source of enduring discord."

Over the years, China consistently denied that it had any intention of building new dams on the Brahmaputra in Tibet. But, faced with satellite photos taken in late 2009 showing construction activities, the Chinese in October 2010 admitted they are building dams, including a 510 MW hydro project, with plans for four more. The *Economic Times of India* reported, "There have been reports that these projects are the beginning of a much bigger plan by China to divert the waters of the Brahmaputra to feed its parched northeast, an ambitious and technically challenging plan, called the Western Canal, that many Chinese reports say will be completed by 2050."

While running only some three hundred kilometers, the Western Canal would present daunting technical, geological, and environmental issues. Building the canal would require blasting out tunnels and aqueducts at high altitudes and subzero temperatures. A 2006 estimate put the cost of the Western Canal at some $37. 5 billion, compared to the $25 billion needed to build the Three Gorges Dam on the Yangtze River.

China has officially denied plans to divert the river. In late 2009, China told the Indian government that reports of the diversion are not "consistent with facts." Indian foreign minister S.M. Krishna told Parliament during a questioning period in April 2010, "In November 2009, the foreign ministry of China clarified that China is a responsible country and would never do anything to undermine any other country's interests." This statement produced amusement among China mavens, noting that China has unilaterally seized Tibet and taken land from India over the years.

Chinese documents undercut the denials. A 2005 book, *Tibet's Waters Will Save China*, argues in favor of diverting Tibet's rivers from India to China. New Delhi's Chellaney observes, "Diversion of the Brahmaputra's water to the parched Yellow River is an idea that China does not discuss in public, because the project implies environmental devastation of India's northeastern plains and eastern Bangladesh, and would thus be akin to a declaration of water war on India and Bangladesh."

The Chinese apparently believe that nuclear geo-engineering would help overcome the technical obstacles to the Brahmaputra project. According to Chellaney, "Chinese desire to divert the Brahmaputra by employing 'peaceful nuclear explosions' to build an underground tunnel through the Himalayas found expression in the international negotiations in Geneva in the mid-1990s on the Comprehensive Test Ban Treaty. China sought unsuccessfully to exempt 'peaceful nuclear explosions' from the CTBT, a pact still not in force."

Under its current leadership, China has emerged as more truculent and triumphalist in its relations with other countries. George Washington University's China scholar, David Shambaugh, has described China as "an increasingly narrow-minded, self-interested, truculent, hyper-nationalist and powerful country."

Leading China's ambitious plans to drain Tibet for the benefit of the ethnic Chinese regions are party chief and state president Hu Jintao and prime minister Wen Jiabao. Hu, who is scheduled to turn power over to a new generation of Chinese communist leaders within the next year or so, is particularly identified with the project to use nuclear bombs to dewater Tibet. He is a hydro engineer, long a traditional occupation among Chinese leaders, and a former governor of what China calls the Tibetan Autonomous Region. but which much of the rest of the world regards as conquered Tibet, brought under Chinese control in 1949 and strengthened in 1959, when Tibetan patriots spirited the young Delai Lama out of the country and into India, where he has led a government in exile.

Wen is a geologist by training, and has a long association with Chinese government projects involving geology and hydro development, including the Three Gorges Dam.

Will China continue with its still undisclosed plans to use "peaceful nuclear explosions" to divert the Brahmaputra River system from its not-entirely-friendly neighbors of India and Bangladesh to water arid Chinese lands? That question is open, and the emergence of Xi Jinping as the likely successor to Hu in 2012 could soften Chinese geo-engineering plans. Fifty-seven-year-old Xi is a lawyer by education, a Marxist theorist, and a veteran bureaucrat. He's often portrayed as a softer figure than either Hu or Wen, although there is little tangible evidence for this view. But the persistence of rumors of a Chinese plan to use nukes to rearrange the earth verifies the observation that dumb ideas die hard.

Breed It

Breeder reactor enthusiasts have always been possessed of a sort of religious fervor. Maybe it's an engineering thing. Engineers seem to venerate efficiency, even when it isn't economically efficient, so the thought of leaving unused energy behind in spent nuclear fuel rubs a lot of nuclear power advocates against the grain.

The notion that conventional nuclear power might be leaving a lot of energy on the table—or buried in a desert somewhere—has motivated a significant number of nuclear advocates to hew to breeder reactors and reprocessing: no matter what. So the death of the Clinch River Breeder Reactor didn't represent the interment of the cult of breeders and plutonium reprocessing.

Not long after the nuclear devastation in Japan following an enormous earthquake and tsunami, an official of the World Nuclear Association in London (the spawn of the uranium cartel) was arguing that the events at Fukushima made the case for closing the fuel cycle. Steve Kidd's reasoning was that because the spent fuel pools at Fukushima were damaged, it would be a good idea to empty them into plutonium reprocessing facilities. But, even if reprocessing were in place—as it is in Japan—those spent fuel pools would have been full of waste, waiting for reprocessing. Never mind.

Kidd even argued, rather astonishingly, that the Fukushima aftermath demonstrated that "the United States is gradually moving away, it seems, from a clear preference for the 'once through' nuclear fuel cycle, with the termination of the Yucca Mountain repository project." Only a true believer could see that vision. Rather, the United States is clearly moving toward permanent, at-reactor storage in large, dry casks.

The Bush administration, late in its eight-year run in Washington, tried to revive reprocessing and breeder reactors, through an ill-designed and ill-fated Global Nuclear Energy Partnership aimed at building a multi-national program to supply the world with new reactors fueled with mixed plutonium and uranium from first-world reactors. It didn't survive the laugh test.

Salt It Away

With Yucca Mountain's waste dump program shut, those who continue to push for burying the byproducts of the U.S. nuclear endeavor were once again looking kindly on salt deposits. In particular, they are looking at a site near Carlsbad, New Mexico, where a small, test project has been storing wastes from the nuclear weapons program for over a decade. It's known as the Waste Isolation Pilot Project, or WIPP.

When President Obama pronounced his death sentence on an already senescent Yucca Mountain nuclear waste project in 2009, he attempted to soften the blow a bit with a classic Washington ploy. He announced appointment of a committee of credentialed and experienced insiders to advise him where to look next for nuclear waste disposal. These sorts of actions—political window dressing—seldom result in concrete results, but are popular among the pols nonetheless.

Out of either naïveté or a sense of humor, Obama named—what Washington has long-termed "blue ribbon panels," to denote their putative quality—the Blue Ribbon Commission on America's Nuclear Future. The commission was part of a compromise between Democratic Senate majority leader Harry Reid of Nevada, the most dedicated opponent of Yucca Mountain in Congress, and the administration. Reid wanted a congressionally-appointed commission to scope out what should come next in waste disposal, but the administration wanted more control. In March 2009, Reid and Energy secretary Steven Chu agreed to a White House–named panel, which the administration announced in January 2010. The chairman of the panel was former Democratic congressman Lee Hamilton of Indiana.[140] Concluding that "[a] new strategy is needed," the commission laid out its views in July 2011, and was unable to come up with any-

---

140 Hamilton was another familiar Washington fixture, the utility policy advisor who serves repeatedly on commissions designed to deflect the politics of controversial issues. Before being named to the nuclear waste commission, he chaired the 9/11 Commission President George W. Bush appointed to offer policy guidance, largely ignored, following the terrorist attacks on New York City and Washington in 2001. Other members of the nuke waste panel included Richard Meserve, a former Nuclear Regulatory Commission chairman, Per Peterson of the University of California–Berkeley, and former Indiana Democratic Rep. Phil Sharp, head of Resources for the Future, a Washington environmental and energy think tank.

thing new. Acknowledging that Screw Nevada had failed, the commission called for new legislation and a new arrangement of the waste management deck chairs at the Department of Energy, to administer a consent-based approach, rather than the political coercion that characterized the 1987 law. The commission said the approach taken to siting the small New Mexico test for disposing of transuranics could be a model for the future.

More explicitly, some in DOE have been focusing on expansion of WIPP for disposing of used civilian reactor fuel. The trade newsletter *Energy Daily* reported in early 2011 that "Energy Department officials, as well as some governors and lawmakers, are warming to the idea of trying to bury some of the nation's high-level waste at DOE's Waste Isolation Pilot Plant in New Mexico."

But there are serious technical obstacles to turning the much smaller WIPP salt-based storage site into a final spent fuel repository. Two Albuquerque, New Mexico, experts—Christopher Timm and Jerry Fox—discussed some of the limits to using WIPP in a September 2011 paper for *Nuclear Energy International* magazine. They also noted that the original decision to build the project created a considerable political uproar, including opposition by the local congressman, several lawsuits, and a twenty-year delay in the project while it could be restructured and reduced in size to meet local political objections.

Despite the failure of the 1986 law, former journalist Luther Carter, along with DOE waste program veteran Lake Barrett and former NRC commissioner Kenneth Rogers, were pushing for resurrection of the Nevada site. In a fall 2010 edition of *Issues in Science and Technology,* the three lobbied the Obama commission to spit in the face of its creators and support continued development of Yucca Mountain. They argued, "Surely this is not the time to abandon the only currently viable option for very long-term geologic retrievable storage of spent fuel, and possibly final disposal."

It should come as no surprise that the commission deliberately refused to take this action, instead issuing a typically anodyne report advocating unspecified changes to make things right. In the meantime, the nuclear regulators have repeatedly judged that it is safe to store spent reactor fuel above the ground and at the site of the operating reactor. That will be the default position on minding the nation's used nuclear fuel, and there appears to be no reason why it won't continue for as long as anyone can predict.

NOTE TO READERS: If you wish to learn about the sources for this book, please connect to the web site www.toodumb.org, where you will find a bibliography and chapter source notes. The web site also includes a bonus chapter on the cons and frauds that have surrounded the quest for fusion energy, a picture gallery with images of many of the people, places, and things mentioned in the book, and the Too Dumb Film Festival, five YouTube videos related to the five sections of the book.

Index

Printed in Great Britain
by Amazon